SEISMIC ON SCREEN
An Introduction to Interactive Interpretation

SEISMIC ON SCREEN
An Introduction to Interactive Interpretation

J. A. Coffeen

PennWell Books

PENNWELL PUBLISHING COMPANY
TULSA, OKLAHOMA

1421 South Sheridan/P.O. Box 1260
Tulsa, Oklahoma 74101

Library of Congress Cataloging-in-Publication Data

Coffeen, J. A.
Seismic on screen : an introduction to interactive interpretation / J. A. Coffeen.
p. cm.
Includes index.
ISBN 0–87814–364–5
1. Petroleum—Prospecting—Data processing. 2. Seismic prospecting—Data processing. 3. Interactive computer systems. I. Title.
TN271.P4C634 1990
551.2'2'0287—dc20 90–44283
CIP

Printed in the United States of America

1 2 3 4 5 94 93 92 91 90

CONTENTS

PREFACE

This book is about interpreting seismic data on computer screens. It is intended for both skilled interpreters who are new to interpreting on computers, and to some extent for people who are novices at interpretation itself. It should be particularly useful to geophysicists who left the industry and are now returning, only to find that a whole new way of interpreting has come along while their backs were turned.

It should also be useful for:

> Geologists and petroleum engineers to get an idea of the capabilities of the geophysicists they work with, and of the interactive systems those geophysicists have available, systems that they may be able to use in their own specialties.
>
> Managers, processors, technicians, and secretaries to get some familiarity with what the geophysicists are doing at those fancy machines with the bright-colored displays.
>
> Students to add practical information to the theory taught in most courses.

For anyone who is new to interactive seismic interpretation it is the book to read as a preliminary to a manual written for a specific interactive interpretation system.

The book does not explain basic seismic interpretation. That is given in different degrees of completeness in two other books. *Seismic Exploration Fundamentals* covers all of seismic exploration for oil—gathering data in the field, processing it, and interpreting the processed data. The other book, *Interpreting Seismic Data,* is entirely about interpretation, and has a workbook that enables a person to learn the techniques of interpretation. This present book concentrates on the ways interpreting is done with a computer. The three form a series about seismic investigations, progressing from the most general to the most specific.

As with the other two books, the seismic contractors have made this one possible by giving of their time, demonstrating their systems, answering questions, and contributing illustrations. Then why don't I mention names of systems in the text? The main reason is the lag between now (when I am writing this) and later (when you are reading it). It is certain that, during that time, new developments will change many things in most of the systems. So it would be a disservice to a company to omit the improvements to its system that it had made in the interim. It would also be a disservice to you, misleading you about the systems.

So the book is about interactive interpretation systems in general. It is necessarily about various functions and the several ways they can be performed, as well as the types of hardware that can do those things. If you don't have an interactive system available, just by reading the book you should get a feel for interactive interpretation of seismic data.

WORDING

One of the main goals of this book is clarity. For that purpose, I've handled some punctuation in ways that aren't according to grammarians' rules, particularly with commas, quote marks, and hyphens.

Commas—used wherever they seem useful to separate parts of thoughts,
Quote marks—put punctuation inside quotes only when it logically goes there,
Hyphens—used in words that could be confusing without them (for instance, re-draw because redraw can at first glance look like it should be pronounced red-raw).

Broke one of my own rules, too. The title of this book uses "seismic" as a noun—something I've resisted for years. When a geologist would suggest that we shoot some seismic, I'd say we should do some geologic too. But when a reviewer (anonymous) suggested the title, it was just too good to pass up.

1

Interactive Systems

Interactive seismic interpretation is interpreting seismic data by interacting with a computer. You do the interpretation at a workstation, where you give commands to the computer and it responds with words and images on one or more television-type screens (Fig. 1–1).

AT THE WORKSTATION

To interpret interactively, sit down at the workstation and pull your chair into a comfortable position for viewing screen displays. After a few preliminaries, give commands to call up a section. You will be asked by words that appear on the screen for the line number, the shot points, etc., that you want displayed.

The part of a section you have selected appears on a screen. With commands chosen in any of several ways, you can have your picking appear as a colored line on the section, like the line you would have

Graphic Science Corp.

Fig. 1–1 Interactive workstation

drawn on a paper section. If the reflection is a good, easily picked one, you may indicate a starting point on it and have the computer pick it from there. If you don't like the choices it makes, you can go back and change them.

And you can have the computer display a map. It can be a base map or a map of the pick you are making, with different reflection times of the horizon shown in different colors. If you are working all the lines of a 3-D survey, the bands fill in with no space between them, so the changes in color look like contours. Sort of a contoured map forms before your eyes.

But of course, to interpret on an interactive system, you'd have to know what the commands were before you could use them efficiently. Learning the commands is like learning any other skill. It takes some time and effort, but rather quickly gets to be second nature to you. And you don't have to learn them all at once, but can learn a few, and interpret with them until you learn more commands.

The physical skills of traditional seismic interpretation are skills you've had most of your life, for instance, shifting sheets of paper, unfolding to find a certain place, coloring. When you learn to interpret on paper sections, those abilities are assumed because you have practiced them most of your life. You can concentrate on learning how to get useful information from seismic reflections.

But the simple skills of using a computer are not skills you learned in childhood. So you have to learn the computer's equivalent of paper-handling in order to interpret by computer. If you already know how to interpret, then you have to learn these new manipulating skills, so you can use your knowledge in this new way. A book on seismic interpretation assumes that you already know how to handle paper, and goes from there in teaching you about interpretation. This book goes the other way around, assuming you know something about interpreting, and explains how to do it on a computer. A generation from now, a similar book may assume the computer skills and just teach the interpretation.

There are some kinds of experience that can help you get started. If you are an experienced interpreter, then a major change will just be to an easier way to interpret. If you have worked in data processing, you have a familiarity with computers that will make it easy to adjust to interactive interpretation. Experience can come from using computer programs to handle geological data. Experience with word processing will make you familiar with giving commands to computers and with the way they don't forgive the slightest deviation. And, of course, if you have played computer games, you have the same familiarity. Any of those types of experience gives you a head start, but even without it, the learning process isn't nearly as difficult as, say, learning to type or to drive.

Why bother with the new way? Because it's much faster, in many ways it's more convenient, and it enables you to do better interpretation. Also, why did it come about? What is causing this major change in the interpretation of seismic data?

DATA EXPLOSIONS

The history of seismic interpretation has been one of tasks becoming easier, and then of harder tasks being taken on. That, of course, is in the nature of progress of all kinds, not just in seismic interpretation. Interactive interpretation has made the mass of data in a 3-D survey manageable, and the interactive systems make you able to do kinds of interpretation you could never do with paper sections.

During the development of the seismic method, there has been a progressive increase in the amount of data handled in interpretation. Some early seismic records had two traces each. Through the years the number of traces on a record increased to six, twelve, and twenty-four. And the number of shots recorded in an area increased as coverage of the subsurface became more detailed. These changes were the general increase that might be expected as refinements in the method.

But there have been two explosive increases in the amounts of data to be interpreted, and each of the two has encouraged the development of a new way of interpreting. In each case, the quantity of information required a more efficient way of interpreting.

OFFSHORE DATA EXPLOSION

The first of the data explosions was caused by offshore shooting. The amount of data that could be collected by a land crew in a month when the operations ran smoothly was suddenly exceeded by the amount an offshore crew could record in an ordinary day. And seismic data was still being recorded, one shot point at a time, on photographic paper. When people continued the old way of interpreting—calculating corrections by hand on each record—the quantity of offshore data swamped them. This was during the time when each crew had its field office, in which the area was interpreted as it was being shot. For offshore work, the field office staffs had to be much larger and more organized in order to do the usual interpretation.

One contract company made a special effort to get a good start in its first offshore operation. Its people asked clients, competitors, anyone they could, about how to run an offshore crew. The result was that to our, and probably their, amazement, they achieved normal production on their first day in the field. But they'd asked people mostly about field work; they assumed they already knew how to run a seismic crew's field office from their years of experience with land crews. They knew the new office would have to be bigger, but didn't know how many employees would be needed. So they rented a large office space, but staffed it with the usual few people, enough to interpret about 100 shot points a month.

After a few weeks, the head office phoned the Party Chief, "How many shot points are you behind now?"

"About 4000."

A dozen additional people were there in a rush.

A better way of handling large masses of data suddenly became economically necessary. About that time seismic data started being recorded on tape. This magnetic recording opened up the new field of data processing. The processors shifted traces and made other corrections on single records. Not long after that they started printing these corrected records side by side in long sections. Interpretation went much faster when the interpreter could mark a section that was already corrected for elevation, weathering, and normal moveout, instead of using the old method of laying individual single-shot records side by side and marking irregularly curved reflections from one to another.

Other improvements, particularly CDP shooting and digital recording, made the quality of the data better and so made interpretation more powerful. With CDP, sections were made up of data blended from different shot points. Digital recording made it possible for processors to greatly improve data. But these improvements did not greatly affect the quantity of data the interpreter worked with.

3-D DATA EXPLOSION

The second explosion in the amount of data for the interpreter to handle was brought about by 3-D shooting. Instead of shooting specific lines that connect with other lines at only a few points, 3-D consists of lines so close together that they form in effect a blanket of coverage of an area. Working a 3-D survey with paper sections took a great amount of time; it also didn't allow the interpreter to make as many comparisons of different sections as that amount of data called for. Again a faster and better way of interpreting was needed because of a great increase in the amount of data gathered.

Some attempts were made to solve the problem mechanically, for instance putting the sections on frames of movie film and projecting them one at a time onto a translucent working surface. But the real solution turned out to be to store the data in a computer and call up sections as images on a screen.

In both these explosions of data the solutions have been so superior to the older ways of handling data that they were adopted for smaller quantities of data also. The sections that offshore data needed were also useful on land, where they weren't needed from a quantity viewpoint, but were used because they permitted better interpretations. The computer method needed for 3-D data is also an improved way of looking at 2-D data.

TRANSITION PERIOD

We are in a transition period in seismic interpretation. The development of 3-D shooting is bringing about the change from interpreting on paper to using interaction with a computer. Before this transition period, all interpretation was done on paper; at some time in the future interpretation will almost all be done on computers.

Anachronisms, the crazy mixtures of old and new that we are often startled by, tend to occur in times of transition, when something new is crowding in on the old. The changes that are taking place in seismic interpretation are bringing an anachronistic blend of methods to it. This is new to interpretation, but not to seismic work as a whole, which, for a different reason, has tended to be full of anachronisms. They occur because it is a highly technical endeavor conducted in all sorts of places.

Once a field crew I worked on had its equipment mounted on mule-drawn wagons because the landowner didn't want trucks making ruts in his land. In a power outage one time we had to use an office copier by candlelight. There was emergency electrical power from a generator to operate the copier, but not to see by. Now someone may interpret an area interactively, using the most advanced techniques available, and then copy the picks onto paper sections to pack in a briefcase and show to someone in a farmout proposal.

The acquisition and processing of seismic data are also in transitions of their own. Their changes are intertwined with the change in interpretation.

ACQUISITION

Data acquisition is changing with the increasing use of 3-D shooting.

The best use of 2-D shooting is in reconnaissance surveys. It is economical to shoot a few widely spaced lines to get a general idea of the geology of an area, and to locate some prospects. Then after interpreting those lines, you might shoot more lines between the first ones to make the spacing between lines smaller, or you might shoot a pattern of closely spaced lines on an anomaly turned up by the first shooting. In most seismic investigations with 2-D lines, the detail on anomalies proceeds in several stages, each time adding more lines. The detail lines may be in a close grid, a set of radial lines, or one or two specific lines to answer specific questions. These last, the one or two detail lines, rarely answer the question they were intended for, as the geology almost always turns out to have complications that weren't thought of when the lines were assigned.

In 2-D shooting, if parallel lines are shot, they are connected by tie lines shot at an angle to the first lines. The first set of lines is usually shot extending up and down the general direction of dip of the rock layers in the area. The tie lines, since they are in the direction of strike and therefore don't have as much dip, don't need to be as close together.

A 3-D area consists of a number of parallel 2-D lines, shot so close together that a new line in another direction can be made up of a trace from each line, forming another line of points like the traces of a regular 2-D line. So tie lines don't have to be shot, they can be made from the other lines. The 3-D lines are normally something like twice as far apart as the traces of one of those lines. Just as in 2-D shooting, you need closer control in the dip direction.

While 2-D is best for reconnaissance, 3-D is primarily for fine detail. It is used for situations that make close, expensive detail worth the price. That gives 3-D some special uses. It is ideal for development work. When an oil and/or gas field is discovered, the amount of money that will be spent on developing it is many times the money that was spent in exploring to find it. Exploration costs must be kept limited, because most of the area explored will not produce.

But a field that has already been discovered is in a totally different category financially. Although some hundreds of thousands of dollars may have been spent to discover it, millions of dollars will be spent on development wells and supporting facilities. The cost is especially great offshore, where several platforms, each costing tens of millions of dollars, may be needed. Putting them in the best locations can get the best benefits from that investment. A 3-D survey over the area provides detail that can be fitted in with production data obtained later.

A further use for 3-D in a developing field is in re-shooting the survey as the field is produced, to determine changes caused by the production and by injection of fluids in some wells. For this use, the geophones may be buried and left permanently in position for all future shootings of the area.

A 3-D survey may also be worth the expense on an undrilled prospect, especially if exact location of an exploratory well is critical to discovering oil. In some exploration situations, the risk of drilling may not be reasonable without a 3-D survey. For instance, if narrow channels are the prospects, then 2-D shooting may not be adequate to map them.

The distinctions between 2-D and 3-D shooting are lessening in several ways. The main use for 3-D has been for detail. But now seismic contractors are shooting areas in 3-D speculatively. They are selling the results either complete as 3-D surveys, or as selections of lines from them, making up 2-D surveys. And oil companies may join together in group shoots to spread the cost among them. In some offshore areas, obtaining data from speculative 3-D or group shoots can be cheaper than reprocessing older 2-D data, so these data-sharing schemes become practical sources of reconnaissance information.

A fairly close 2-D survey that was shot on a regular grid may be made into a sort of 3-D survey by interpolating between the lines to create synthetic lines. This type of interpolation is also used on 3-D data to avoid aliasing in migration.

Some areas are being shot with a low degree of CDP, by just shooting two perpendicular lines while recording both lines simultaneously. This makes an economical 3-D, shot in an almost 2-D manner.

How soon will 2-D shooting be a thing of the past?

DATA PROCESSING

Data processing is just started in a transition from being done only by processing specialists in a specialized setting, to a stage in which some of it is being done by interpreters at their workstations.

Amplitude on a paper section is set in playback, and to get a section with a higher or lower amplitude setting would call for having a new playback made at the processing center. But in an interactive system, you set the amplitude of the display as often as you like. Other parameters you can set that formerly had to be done in processing are polarity of the traces, scale, and ratio of horizontal to vertical scale. And you can get displays of seismic attributes (Fig. 1–2).

There is considerable variation in the amount of processing that different systems can do. In some you may be able to perform all the steps of processing. It would take too long for you to actually do the processing on a whole area, but the ability lets you try out processing on parts of lines, to see how it could best be done in regular batch processing. And during interpretation it gives you the power to go back and try different processing to see if that would solve some interpreting problem.

Wow! Can doing your own stacking and migrating be far off? Scary, isn't it? Are you to be expected to be a skillful processor as well as interpreter?

INTERPRETATION

And in this transition period, you are likely to use paper sections right along with an interactive interpretation. The clues are everywhere—in ads, in articles, in talk sessions—many seismic interpreters using the new way of interpreting are also using the old way along with it. There must be some good reasons for using paper sections and maps while also using an interactive interpretation system.

If interactive interpretation is so much quicker and so much more accurate, why bother using paper along with it?

Some of the main reasons seem to be:

1. Interpreters, both those who are new to computers and those who are familiar with them, may be concerned about losing information that is stored in the system. The basic data, the traces that are loaded into a system, are fairly safe from loss but the picks you've made are subject to accidental deletion. They have to be erasable so you can change your mind while interpreting. So there is always the possibility of giving a wrong command and erasing picks you wanted to keep. Of course that is the reason for backup copies, to avoid losing much of your picking, but you may not make backups often enough, or something may happen to them. It's reassuring to have your picks safely marked on paper sections.
2. Most people feel more comfortable doing familiar things. Most seismic interpreters probably have developed methods of work that don't necessarily fit in with using a computer. They feel at home with paper sections.
3. In any transition like the one seismic interpretation is going through now, it is safer while learning a new technique to also have your work in a familiar form. You know the traps in the old method: misplacing sections, grabbing the wrong copy, etc. But you may not yet be alert to all the traps in the new method: deleting the wrong file, failing to make frequent backups, etc.

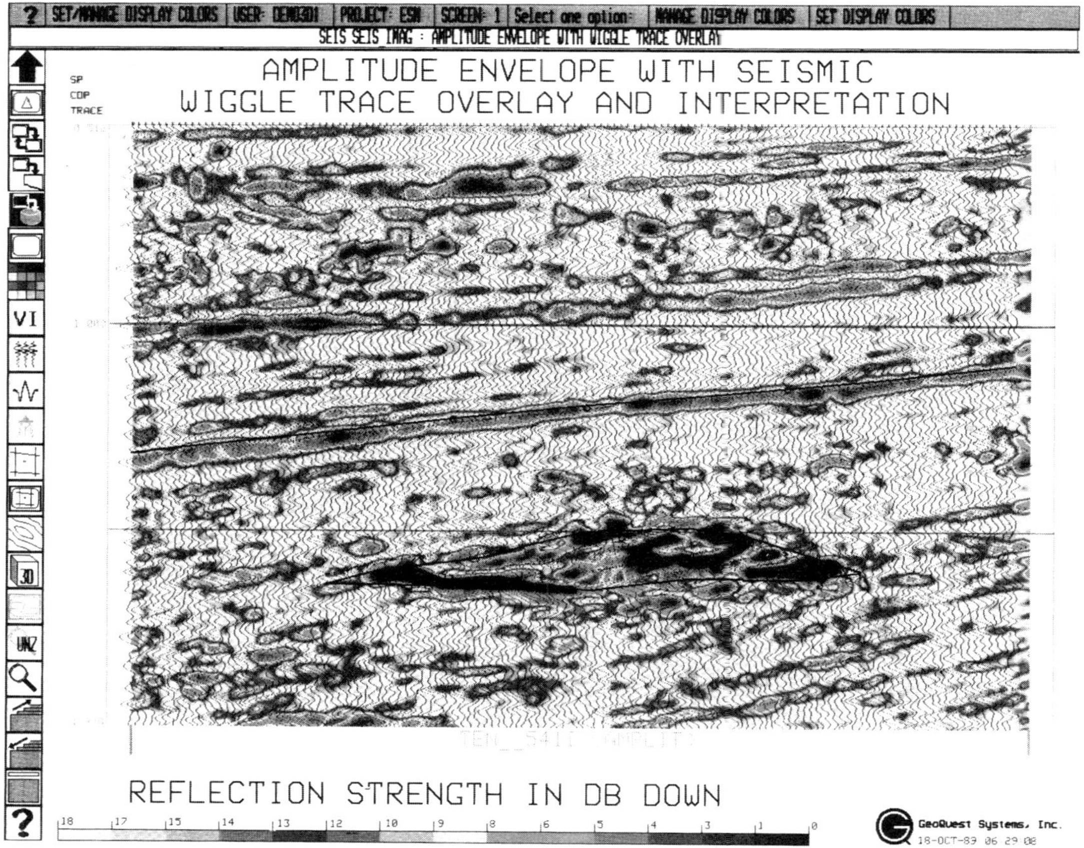

GeoQuest Systems, Inc.

Fig. 1–2 Attribute display

4. A computer screen doesn't fit certain situations in seismic interpretation very well. Some 2-D lines are too long to be displayed all at once on a screen. Some areas are too large to be shown on a screen in map form with a useful degree of detail.

Items two and three fit what could be called the buggy-whip syndrome. Some of the first automobiles had sockets for buggy whips. Perhaps they were there because people were accustomed to having whips handy for the familiar horse-drawn vehicles. And maybe there was a small practical side to it too, that cars often broke down or got stuck in mud, and had to be pulled by horses. In that case a buggy whip might be handy. In the same way, in interactive interpretation, we like our buggy whips both for reassurance and for the occasional practical use.

Here are some situations in which both reassurance and occasional practical use apply:

1. As you interpret on the computer, if you keep one or two main horizons colored up-to-date on at least some paper sections and a contoured map, then you have something to show your boss, take to a meeting, in case "the computer is down" just when some important decisions are to be made about the area.
2. When the equipment is not available, because someone else is using it, or it is being serviced, you can make some progress on the paper sections.
3. When interpreting a long line interactively, it may be useful to also have a long paper section of that line on the wall or on a desk.

So far, the interactive systems are expensive and the ones already bought are in such demand for 3-D that most 2-D interpretation is still done by hand. Apparently only a small percentage of 2-D interpretation is being done interactively at present. That will change with time, as more systems are bought, and as they get cheaper (because of advances in their design and economies of scale in producing greater numbers of them).

How soon will interpreting by hand be something for old-timers (timers, hah!) to reminisce about?

INTERACTIVE INTERPRETATION SYSTEMS

An interactive interpretation system consists of the hardware and software, and is controlled by a person giving simple commands. It performs seismic functions, like displaying sections and storing picks. Since the capabilities and ways of executing them differ from system to system, and within a system as time passes and improvements are made, the book will treat the systems in general, not specifying how a particular system performs a certain function or even which system does so.

FUNCTIONS, WHAT THE SYSTEM DOES FOR YOU

Here's this system facing you. It's big, it's electronic, and it's expensive. It's also the current rage, and you get the impression that if you want to be up to date, you need to know how to use it. That's great if you happen to want to be up with everybody, but you are more likely to just want to do effective interpreting. Will it do anything to improve your interpretation? Being modern is good, but it had better be useful as well if it's to be worth the expense of acquiring it and the effort of learning to operate it.

Just what are the functions of interactive interpretation systems? What do the systems do? A company that produces a system has to put its main effort into designing functions that its competitors don't have, and producing functions similar to ones the competition already has. Its programmers are kept busy developing functions. Functions sell systems. Potential customers ask if a system has certain functions. Functions are what I am describing in most of this book. So far in the book, some of them have been mentioned, either directly or indirectly.

Once, in the days of just interpreting on paper, I had a boss who said we needed a nine-year-old kid to help us. His point was that most of our time that should have been spent on interpreting was instead spent on tasks that a kid could do, tasks like looking for things and putting them away again. That is much of what the computer does, the things the little kid could do. But it does them faster, and it always has everything available. It gets sections out for you; it files them away when you're through with them. And it does things that the kid couldn't do. It reads the times of picks; it keeps the times in a list; it plots the times on maps; it colors things like you want; it transfers your picks from one form of a section to another. It makes your rough picks more correct. It can sometimes do its own picking. It displays various seismic attributes. It shifts parts of sections, flattens sections, calculates intervals between horizons, and converts time to depth. And of course it doesn't get tired or bored, and doesn't get into mischief when it is idle.

There are comparisons you couldn't make at all on paper that are easy with interactive interpreting. In a 3-D block of data, the computer can show you section after section, marching through the mass, so you can keep your eye on one spot to see how it changes.

To really have the paper sections you need, you would have to have several of each line: one large scale, one small; one in a height-to-width proportion that is easy to pick, one highly compressed horizontally; one with the stronger reflections outstanding, one with reflections nearly equal, etc. The filing problem alone makes this impractical with paper sections. And the cost of making all those sections would be prohibitive. Making a variety of sections on paper, especially color sections, would cost a great deal of money. For a large 3-D survey, even a set of ordinary sections, without color or alternate displays, could cost a considerable fraction of the price of a system. With an interactive system, seeing plenty of sections is easy.

Consider how much work is involved in interpreting a horizon on a single seismic line on paper. You first identify horizons to pick and then mark your picks with colored pencil. So far, you haven't done anything much more laborious than you would on an interactive system. But, having marked picks, you then have to time them by determining the second of reflection time the pick is in, counting tenth-of-a-second lines, and perhaps estimating closer than that. You have to write the times either on the section, somewhere near the reflection, so the traces get in the way and make your numbers hard to read, or in the white space at the top of the section (but that gets confusing if you also put the times to other horizons up there). Or you can list them with the shot point numbers on a separate sheet of paper, being sure to have the correct line number on the sheet. Now you have the reflection times. Then they must be posted on a map. After the numbers have been through all those steps, there is a good chance that some errors have been made. So someone should check them, preferably checking both your reading of times and your, or someone else's, copying of those times onto the map. That's a lot of work, it takes a lot of time, and after all the checking there can still be some errors remaining.

On an interactive system, as soon as you've marked your picks, you are already through with those chores of timing and posting that took so long on paper. The computer reads the times as fast as you mark them, files them away for future use, and can put them on a map. And the work is correct.

The functions either are performed automatically or commands must be given for them to be performed. In learning to use a system, you will be largely memorizing commands. But the commands are given so the system will perform functions. So before learning commands, you need to learn what functions are available in the system, or rather, the main functions, the ones you will be using when you start on the system. Then, knowing what is available, you can learn how to tell it to do those things.

The standard functions tend to have different names on different systems, so you may have trouble talking to a person who has a different brand of system from yours. The different names do serve a purpose for the vendors. If a function on one system actually is somewhat different from other versions, then giving it the same name as the others have, would fail to notify potential customers that there was a difference.

However, the functions fall into a few main categories:

Filing and finding. The system keeps your sections filed for you. You never have to put the sections away, but you can find any one whenever you want it by "calling it up" with a command. That's a big step there, considering how much time is spent in digging paper sections out and putting them away again. A system performs the functions that caused it to be invented—handling the mass of data in a 3-D volume so you can work on it effectively.

Displaying the data in different forms. It shows you a section in any of a number of different forms: different colors, different trace displays, different scales and proportions, seismic attributes. The various forms can be shown on time slices as well as on vertical sections. Traces can be shifted up and down to flatten horizons and to adjust the times of reflections on one line to those on another. It is as though you had a tremendous file of varieties of sections and slices.

It can show the different forms in combination displays: sections connected where they meet, section and time slice together, and solid-looking displays like cubes (Fig. 1–3).

Interpretation is visual. If a system makes some feature of the data easier to see, that's effectively an improvement of the data.

Picking. It helps you in picking horizons and faults. You can pick horizons roughly and have it move your picks to exact peaks, troughs, or zero crossings. With good reflections you can have it do the picking automatically, then you can glance at its picks, and, if it went wrong, correct the picking easily. If its exact picks are jagged-looking, you can have it smooth them, doing much the same thing you do when you smooth by eye as you pick.

You can mark breaks in reflections where there are faults, and the system will determine the width and throw of the faults. It will display faults in perspective to help you decide how to connect them.

When you want picks erased, it removes them completely. That's because it files the picks separately from the section. It's as though you picked paper sections on transparent overlays, and then, when you changed your mind about a pick, just threw away the overlay and got out a new one.

Timing and posting. It reads the reflection times of your picks automatically, instantly, and correctly. It saves those times and can post them on a map for you.

Keeping track. A system not only keeps track of the sections so you can have one whenever you want it, but it also does wonderfully at keeping track of your picks.

It keeps track of the picks on tie lines at line intersections, showing the position of the latest pick from the intersecting line, no matter how

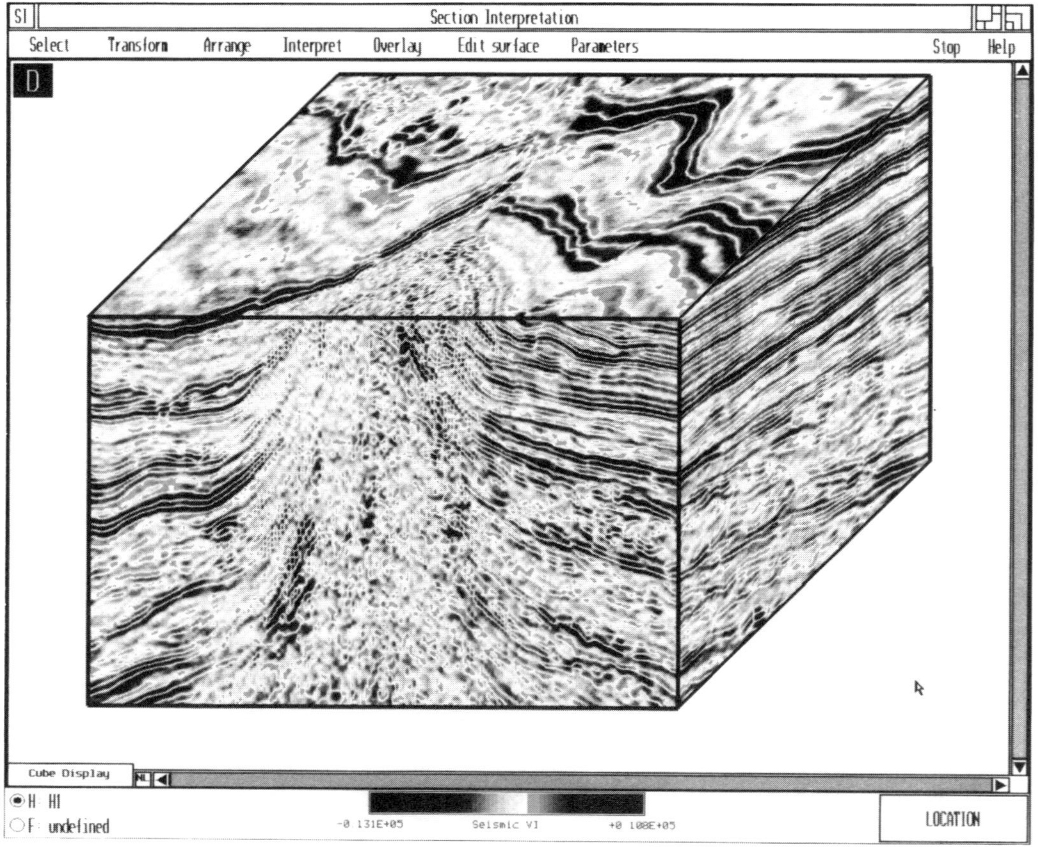

GeoQuest Systems, Inc.

Fig. 1–3 Cube display

many times you change picks. And, no matter what form of section you picked on, it can show those picks on any other form you wish.

Making calculations. With your picks in memory, the system can perform calculations on them, like converting times to depth or subtracting one horizon from another.

Mapping. The system shows you where you are on a map; you can select lines to interpret by pointing them out on the map; the reflection times of your picks are shown in color on the map. The system can contour your picks quickly and correctly, although you may have to adjust the contours to make them geologically reasonable. You can have different kinds of map information (political, lease, etc.), and show any combination of them on the map.

Studies of the data. A system can make a number of investigations of the data, for example determining the average wavelet from seismic data, making synthetic seismograms, and modeling interpreted horizons.

If you have a difficult decision to make in interpreting, it will obligingly let you make different versions in less time than it would have taken you to prepare one version without its help.

Bringing in other information. It can put synthetic seismograms and well logs in the sections, with the logs converted to time, so you can truly use geological information in your seismic interpretation (Fig. 1–4).

Making copies. With a plotter, it can make a paper or film printout of any section or other display. With a special camera, it can make slides of the screen displays, or you can take pictures of the images on screen with an ordinary camera.

An interactive interpretation system is pretty nearly an ideal helper. About the only thing that keeps it from being a perfect helper is the communication problem. That is, sometimes you can't think of how to tell it what you want, and sometimes when you think you have told it, it misunderstands.

HARDWARE, THE EQUIPMENT

At first glance, different interactive interpreting systems look pretty much alike. The first thing you notice is a desk with one or more television-like screens on it. Two screens, for sections and for mapping, can have the main displays. If there is a third screen, it displays lists of choices you can make. If there is one screen, the various displays can be shown alternately. People tend to call a system with three screens a two-screen system because the interpretation is done on two screens. When you look closer, you see that the system has a keyboard and probably a small pointing device: a trackball, mouse, puck, or stylus.

The computer that runs the system may or may not be evident. It may be under the desk, somewhere else in the room, or out of sight in another room. There are several ways the computer can be related to the

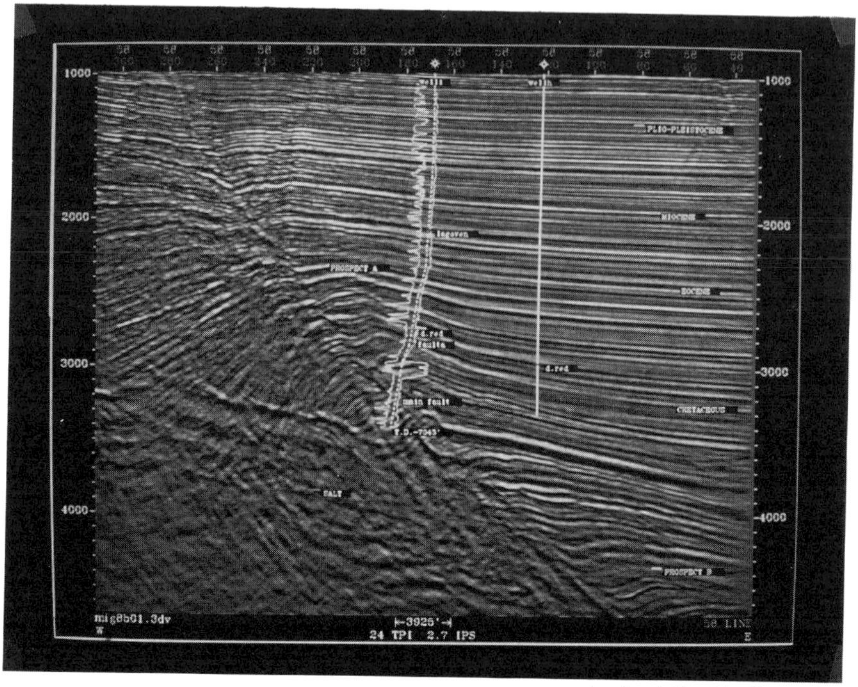

Landmark Graphics Corporation

Fig. 1–4 Well logs added to section

system. The interactive interpretation system may be just the workstation and some software, run off the company's computer, and connected to it by wires through the walls or under the floor. That computer is probably a mainframe, a huge thing that, with its peripherals, takes up a large room. A number of terminals are scattered about, mostly in other rooms, connected to the computer by wire, all using it by timesharing. Each terminal consists of a keyboard and a screen, with some other equipment.

In timesharing, the computer spends a few milliseconds on one project, then does a bit of computing on another project and so on. It gets back to the first program so quickly that the user doesn't notice that it's been away. The projects it works on are of all kinds, like accounting, seismic, payroll, and reserves calculations. A mainframe computer is so expensive that failure to use it to capacity is a significant waste of money. So it is used by many people at their many terminals, and at night when people aren't using it directly, it is kept busy running large programs that need a lot of computer time.

A smaller company or a smaller office of a large company may use a minicomputer instead of a mainframe. A minicomputer is so named because it is considerably smaller than a mainframe. Advances in computer technology are so rapid that a new mini may be more powerful than an old mainframe, so the difference is as much one of size as of power. A minicomputer may take up as much room as a few filing cabinets. It may look like filing cabinets, too, except for the knobs and indicator lights on it. A minicomputer can also handle a number of workstations. But, being considerably cheaper than a mainframe, it doesn't have to be kept as constantly busy; a company can afford for it to be idle part of the time. So it doesn't necessarily have many workstations.

Another way the computer may be related to the system is for the system to have its own computer, so it doesn't have to do timesharing on the company's business computer. Then it is a standalone system. It doesn't depend on another computer. The computer for a standalone system may be so small it is built into the desk of the workstation. That type of computer is somewhere between a minicomputer and a microcomputer (personal computer) in size and capability. Or the standalone system may use a minicomputer, maybe installed in another room.

Whatever kind of computer you have, there is a CPU, Central Processing Unit. It is the "brain" of the computer. There is also a disk drive and there may be a tape drive. The data is stored on a hard disk, a stack of hard disks, or an optical disk. And there may be peripherals, such as printers or plotters to put some of the results of an interpretation on paper, and cameras to put the results on film.

But the various differences aren't apparent when you first see it. And to a person working in front of a screen, the difference between being one of the users of a large computer and being the only user of a small one isn't obvious.

That's it. The setup is enough to allow much faster and better interpretations than can be made with paper sections and colored pencils.

The system may belong to the oil company, or it may be leased from the manufacturer, and installed in the oil company's office. It may belong to a contractor, and be used for the interpretations by the contractor or by other people who pay for time on it.

SOFTWARE, THE PROGRAMS

That first look at the hardware can be interesting if you are just touring the office, exciting if you enjoy doing things with computers, or forbidding if you are going to be expected to use it and don't feel confident about it.

The first and second of those situations are no problem, but the third can be a source of concern. All that hardware is sitting there, challenging you to tell it to do something. Much of the telling is taken care of for you in the form of the interpretation programs that have been written and stored in the computer, probably on a hard disk.

Any computer program is a highly detailed list of extremely simple instructions to a computer. A program that causes the computer to do very much must be a long list of such minute instructions. A program for as broad an activity as seismic interpretation is an extremely long list. The writers of the program are skilled specialists. Fortunately for interpreters, these specialists have made comprehensive programs that only need simple instructions (commands) to cause the computer to do complicated things like displaying a section on a screen or producing a contoured map on demand.

The programs are worked out to do these things well, so they don't suddenly show a section upside down or something equally foolish. Thought has also gone into making the giving of commands simple and natural-feeling, so the programs are "user-friendly," that is, easy to learn and use. They have come a long way in that respect.

But the program is also designed to be powerful, to do a great many things. And doing more things means that there has to be a way for you to tell it which of those things you want it to do, so there are necessarily more commands to be learned. The user-friendliness and the power work in opposite directions.

As I mentioned, the original reason interactive interpretation was developed was to handle the mass of data in a 3-D survey. So the first programs were designed to handle the problems of 3-D data.

But ease of handling, ability to vary displays, and speed of interpretation are advantages that can be applied to any sort of seismic interpretation. It became evident that an interactive system could do special things for 2-D just as it had for 3-D, and that 2-D has some of its own special requirements. So some programs were developed just for 2-D.

And the special capabilities of 2-D programs were added to some 3-D programs so they became capable of handling both kinds of data efficiently. A particular advantage of such a combination system is that it can combine 2-D and 3-D data in one interpretation.

These programs are designed to perform the functions needed for interpreting seismic data. There are other functions that may be included in a system, for instance putting well logs in a form to fit in with seismic data, and modeling to check interpretations. Some interpretation systems have data-processing programs accessible, so you can try out different processing on samples of your data.

You may hear operating systems mentioned. An operating system is a more basic program that works beneath the interactive interpretation program. The interactive interpretation program gives instructions to it and it translates them into simpler instructions for the computer itself. In working on a system, you will work with the interpretation program and will not usually be aware of the operating system.

COMMANDS, THE INSTRUCTIONS

The commands you give a computer are just instructions to it, the way you tell it what you want it to do. You have to learn them, like you have to learn the techniques for any other new skill.

A computer insists that the commands be given just so. The computer is picky, far more picky than a grammarian or a lawyer. Any slightest deviation causes it to balk or to do something different from what you thought you were telling it to do. Then you feel lost, with something strange on the screen, and no way that you know to get back to where you were. These problems are what the drive for user friendliness is all about.

Sometimes the computer will help you decide what to do by giving you a multiple-choice list of commands to choose from (Fig. 1–5). And when you are first starting to use interactive interpretation, there will probably be someone nearby, from either your company or the equipment manufacturer, to help if things go wrong.

Most of the basic work of interpreting has by now been made fairly simple, fairly easy to understand and remember. You move a mouse or other device to place a "cursor" on the screen to point to something on a "menu", a list of commands. The cursor is an image on the screen: a plus sign, an arrow, a pointing hand, or some other shape. You pick a horizon, perhaps by moving a cursor along the reflection. You select a section to be displayed by using a cursor to point it out on a map. You may select colors by pointing to them on a display with a cursor. For more abstract choices the screen may display multiple choices in words, for you to point to your selection.

But there are other things you may want the system to do that aren't so basic to its operation. You may want a printout. There may be a simple command to make hardcopy, maybe clicking on a choice. Or for some plotters, you may have to give a series of commands on the keyboard. If you don't use them often, you may need to keep them handy on a piece of paper.

Learning a new technique involves confusion, mistakes, and problems. The learning is worth the trouble, but during the novice period it can be frustrating and annoying. The confusion in learning a technique on a computer is mostly in the commands that must be learned. There is a way around the confusion problem. At first, you can just learn a few commands, so you can have the system do a few things for you. Then as you use it more, you can learn some new command, then another, and so on.

However, there will be times, especially when you are new to a system, when you'll have to do, not what you want to, but what you are able to. It's like using a foreign language that you know just a little of, or dealing with someone who doesn't know your language well. In either case, the actual things you do or have a person do, are not what you

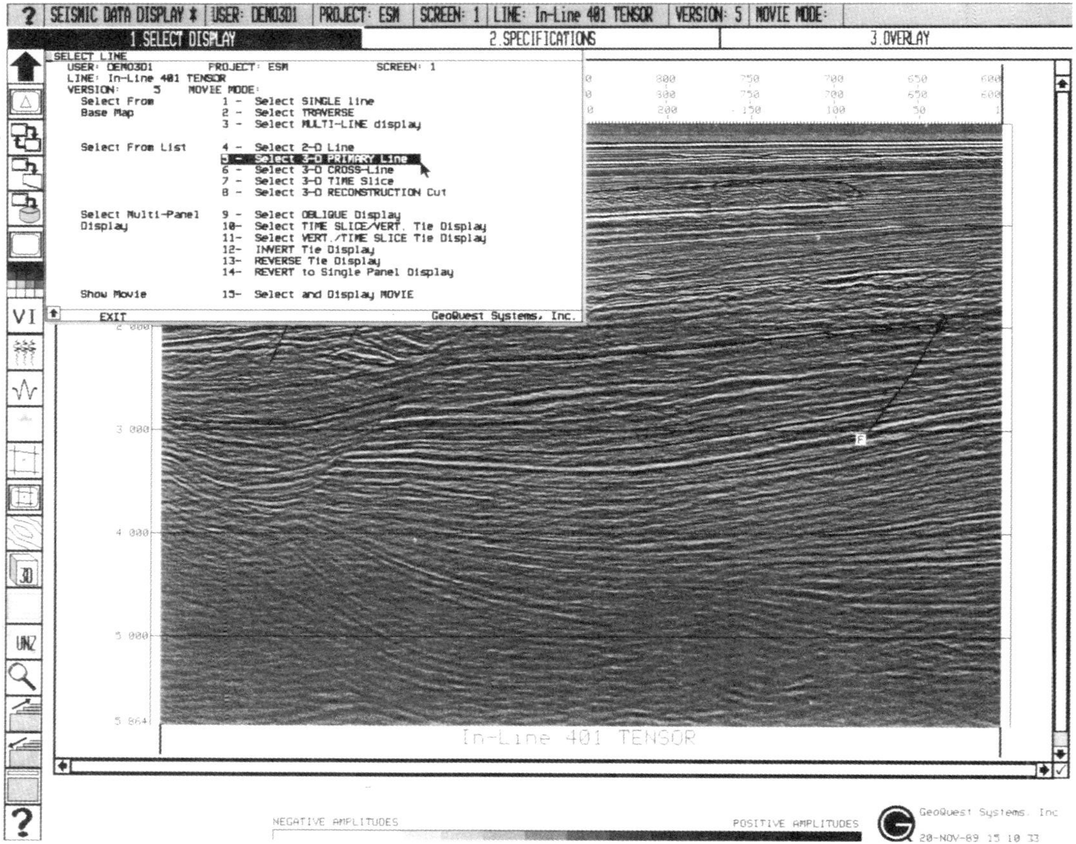

GeoQuest Systems, Inc.

Fig. 1–5 Multiple-choice menu

want, but what you can communicate that is reasonably close to what you want.

But after a while, you'll find that you can make the system jump through hoops for you, and that you can show a newcomer how to get started.

DATA FOR THE SYSTEM

For interactive interpretation to work, of course, there has to be data to work on. A variety of types of data must be entered into the system in order for the interpretation to be accomplished.

The data to be stored in memory is a great amount of information. Any system, whether it uses a terminal connected to a mainframe or minicomputer, or is a standalone system, must have a great deal of memory available. This is especially true for 3-D interpretation. The memory requirements for 2-D interpretation are large, although not nearly as large as for 3-D. In 2-D as in 3-D, every sample of every trace of every line must be stored. The traces in 2-D are just as long as in 3-D, and the lines may be much longer, but there aren't nearly so many lines.

The seismic data in digital form, as numbers representing the changing amplitudes of a trace, extends down the entire length of the trace. The traces are sampled in this way typically once every four milliseconds, but may be sampled every two milliseconds, or even every millisecond. The traces are related to the location data, so each trace can be called up by its location. There are many traces in a line. The number of samples in a survey is enormous. They must not only all be in memory, but in memory in a way that they can be called up at will, and quickly.

The data is put into the system usually from reels of magnetic tape produced by the processing center. These are the tapes that paper sections are made from for manual interpretation. In a large area, the seismic data may take up several reels of tape. But to be used freely in the interactive system, the seismic data must be in one unit, with any part of it quickly available to the computer. Data that is on tape can be accessed only sequentially, by reeling the tape to that particular information. The tape is so long that minutes may pass while it reaches a particular bit of information.

So the data can't be used on tape, but must be transferred to the interactive system, to a random-access form. That is, the computer must have access to any information at random from anywhere in the data. A disk allows quick access to any point on it. The disk spins rapidly and a sensor can move in and out to bring it to any point quickly. The information may be on an optical disk that can be removed from the computer (so other disks can be put in, and other areas interpreted). Or it can be on a hard disk that is built into the computer (from which it can eventually be erased to make room for other areas). In either case, the data for the whole area is available quickly.

The seismic data has been stacked to reduce multiples and other noise. It has been deconvolved to sharpen its definition. If it is 3-D data, it has been migrated in three dimensions to put the reflections in their correct positions. If it is 2-D data, it has been migrated along the lines. For 2-D data it may be necessary to have both migrated and unmigrated versions in the computer.

The seismic data is what is going to be interpreted, of course, so it is the main thing you think about getting into the system. But there is other essential data too.

The data for making the base map must be put into the system. In the field the area was surveyed. The surveying produces a mass of data about location, and about elevation of the land or depth of the sea. That information was processed in a computer to adjust discrepancies between readings, to put shot point locations in map coordinate form, and to put depths or elevations in meters or feet. The information was put on magnetic tape, usually a smaller reel of tape than seismic data is put on.

The location data is loaded into the interactive interpretation computer. During interpretation the computer can display locations of lines and shot points, and it can place reflection-time data at those locations, to show contour maps. The elevation data is used in the processing stage, but it may also be put into the system, so a map of topography or a water-depth map can be produced.

If there are wells in the area, their locations are put on the disk, so the wells can be shown on maps. The angularity of deviated wells is entered, so the system can show them at their correct positions at any level. Well log data is also put onto the disk, so synthetic seismograms can be made, and various logs can be displayed. A synthetic seismogram's vertical scale is in time, so it can be matched to the scale of a section. From it, other logs can be converted to seismic time. It is particularly useful to have formation tops and fault cuts, as picked by a geologist, indicated on the well data.

DISPLAYS ON THE SCREENS

Seismic data can be displayed on a screen as sections, combinations of sections, time slices from 3-D data, or sections and slices combined. Picks can be displayed in map form. Those displays can be made permanent with one kind or another of plotter or camera.

A computer display is usually in color. Since the computer is equipped with a color monitor (screen), there isn't any extra cost or effort in displaying in color. The colors can be used to show more features than one color alone could convey. You can change the colors to make the best display for you.

A section displayed on a computer screen doesn't look much like a traditional paper section. For economy, most paper sections are in one color, usually black, on a white background. And a paper section is usually of an entire line, taking in vertically from the top of the section to as deep as you might be interested in. But when you look at it on a desk in front of you, you see a much more limited part of it, the part in easy range of vision. To see other parts of it, you move your head or shift the section around on the desk.

On an interactive system, the display is, of course, limited to the area of the screen. So, to keep a large enough scale to interpret easily, you will probably show a short horizontal extent of the section, and a vertical extent that may not even include the top of the section. That is something like the area of a paper section in front of you on your desk, the area within your view. To look at other parts of the section, you just have the computer display them instead of the part it was showing.

In 2-D data, sections from the lines that were shot can be displayed, and also combination sections. Where two lines meet, the display can show the two at the intersection, like two paper sections placed together at the intersection. Where several lines form a loop, the whole loop can be displayed.

In 3-D, the same displays of lines shot can be shown. These lines are spoken of as inlines. Lines perpendicular to them can be made up of one trace from each inline. Such a perpendicular line is called a crossline, or xline. Sections can be made up of traces from different inlines to make lines in any direction, including zigzagging across the area.

Another type of display that essentially belongs to 3-D data is a time slice (Fig. 1–6). It is like a section, but in a horizontal direction. It is made by displaying one sample of each trace in the area, all from the same reflection time, for instance showing the samples at a time of 2.004 seconds from each trace on each line. It looks like, but isn't the same as, a contour map. A time slice has lines of reflections that curve around like contours on a map, but it differs from a contoured map by showing contours of different horizons at the same time, instead of the different time contours on the same horizon that a contour map shows. A time slice, like a section, is usually in color.

These special 3-D displays don't apply much to 2-D. With 2-D data, there is no particular need to show one trace from each line, because the lines are so far apart that those traces wouldn't make up a reasonable new line. And a time slice wouldn't usually make much sense, either. It would look like wide-mesh lace, instead of a solid blanket.

An interactive interpretation system can display sections and slices

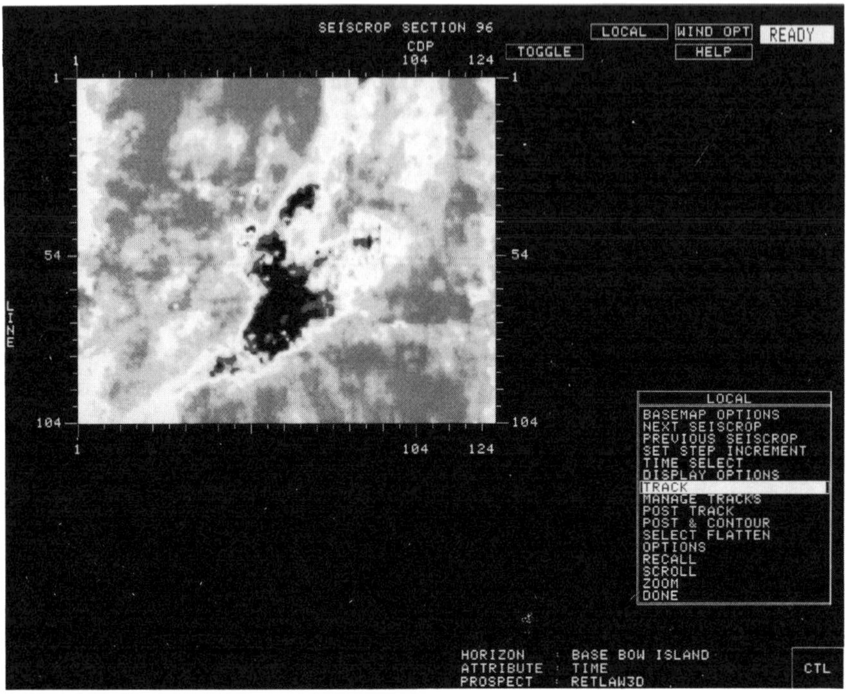

Sierra Geophysics, Inc.

Fig. 1–6 Time slice

in various combinations. Two sections can be shown joined at their intersection (Fig. 1–7). Several such joins can make a crooked line of sections, for instance to make up a section through the locations of several wells. A section and a slice can be joined where they intersect. And several joins can be made on one display and shown in perspective, to give the impression of a look into a corner, or from outside a cube.

All these displays can show any of several characteristics of the seismic data. They can show the traces as amplitude in variable density, in variable area, in wiggles, or in a combination of those. Or, usually as variable density, they can show reflection strength, phase, or some other of the seismic attributes.

A base map can show where the lines and shot points are. The results of interpretation can be displayed on a map in which colors appear like contours, or can be shown as contour lines on a true contour map. Color can be used on a map to distinguish between seismic data, shot point locations, coastlines, etc.

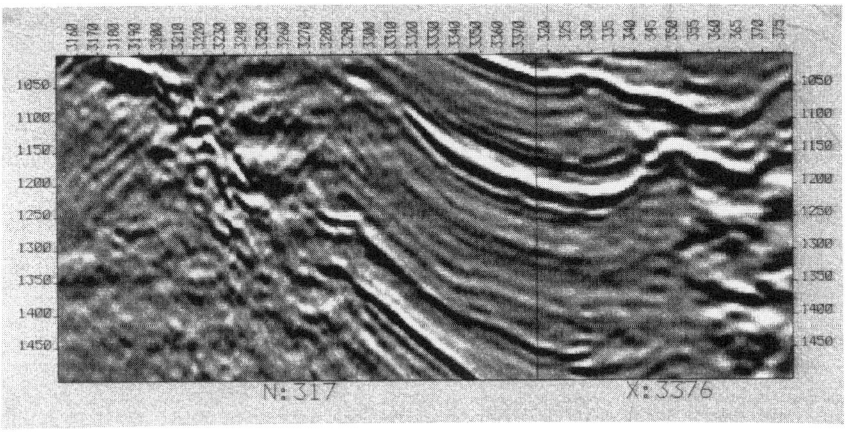

Western Geophysical Division, Western Atlas International, Inc.

Fig. 1–7 Two sections joined

DISPLAYS CAPTURED ON PAPER AND FILM

Interpreting on screens is fine, it's a good example of the paperless office we keep hearing is coming. But that paperless office isn't here yet. We still need physical copies to carry around, show to people, and put in reports. There are several ways to copy the screen images to hardcopy, either on film or paper, from the information in the computer.

Interactive interpretation systems often have color plotters connected to them. If you have one connected, a simple command will cause a section, slice, or map to be printed out. The plotters are slow. Some put down one primary color at a time until the picture is complete in that color, then print a second color over it, and then the third color, so there is finally a good, bright copy of your screen image. The plotter may be limited in the scales of copies it will make. Instead of being able to produce any scale you choose, to fit a report or something, you may be able to plot at one fixed scale, and twice that scale, and then twice that one. Or the plotter may be thoroughly adjustable between two extremes of scale. A copy made on a color plotter is usually sharper than the screen image you "copied" it from. That is because the plotter image is made up of more dots per inch than the screen image has. The copying is really from the computer's memory rather than from the screen.

The paper used in the color plotters is usually of a special type, shiny and easily crinkled, not the best for reports. But the copies are available at any stage in your interpretation, not only sections picked as of the moment, but up-to-date pseudo-contoured maps.

Imagine having that without a computer. The nearest approach to that ideal that I know of was used on field crews long ago, when we did interpretation in the field office. Once a week we sent in a progress report, consisting of a written report on the accomplishments and problems of the week, with contoured maps of one or more horizons. Getting a map ready every week was a chore. It was kept up to date on a paper transparency. Seismic times were drafted in ink. Contouring was done in pencil, so it could be erased and changed when more data was in for the next week's report.

Well, with an interactive system and 3-D data, you can have a pretty good map any time you want. Technically it isn't contoured, but the colors show something like contours, with data in a certain range of seismic times in one color, the next set of times in another, and so on (Fig. 1–8). And for 3-D data, the color bands on a line are wide enough to touch the colors on the next line. So the solid mass of color bands looks like a

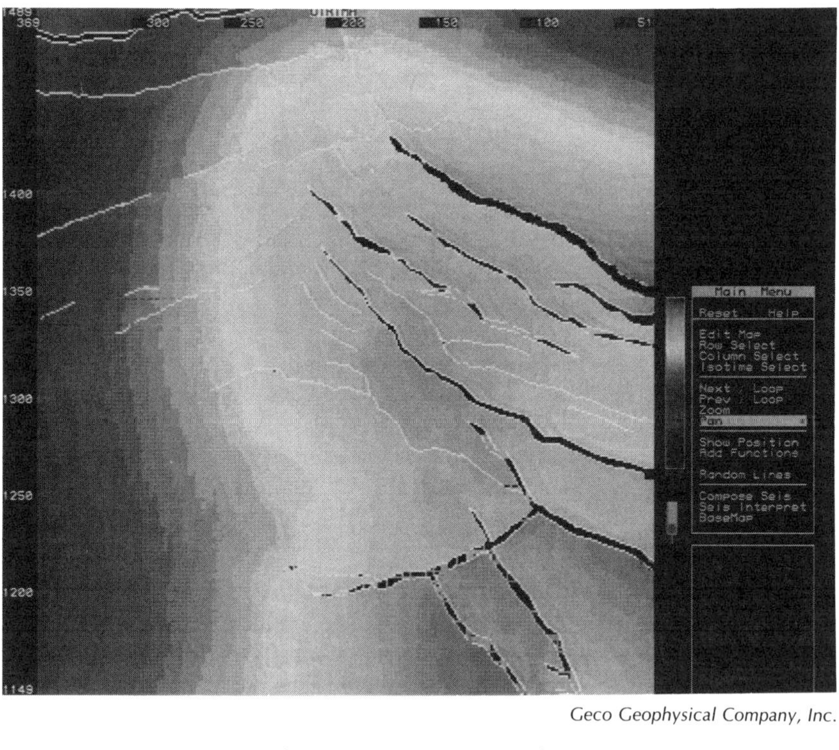

Geco Geophysical Company, Inc.

Fig. 1–8 Color as contours (see in color on cover)

contour map, with the boundaries between colors serving as contours. Prints of that type of map can be made as often as you want.

If you are working with 2-D data, the color map isn't so much like a contoured map. Each line you've picked has a band of color corresponding to the reflection times of the horizon. But the lines are farther apart, so there are gaps between the bands of color. But at least that map is available. Some systems can draw actual contours quickly, so you can have a contoured map whenever you need one.

The copies made on plotters can be generated in several ways. The simplest is a "screen dump", a copy on paper of whatever happens to be on the screen. Your printout of a section will also include any boxes containing commands that are on the screen. You can get it easily, maybe by just pressing one button. Most of the illustrations in this book are made from screen dumps, sometimes by cutting the paper prints to show a detail, other times the entire screen image, commands and all (Fig. 1–9).

To make an image that is better for reports, you can use a function for saving images. The system may call them slides, pictures, frames, or something. You give the command for it, then form a rectangle around

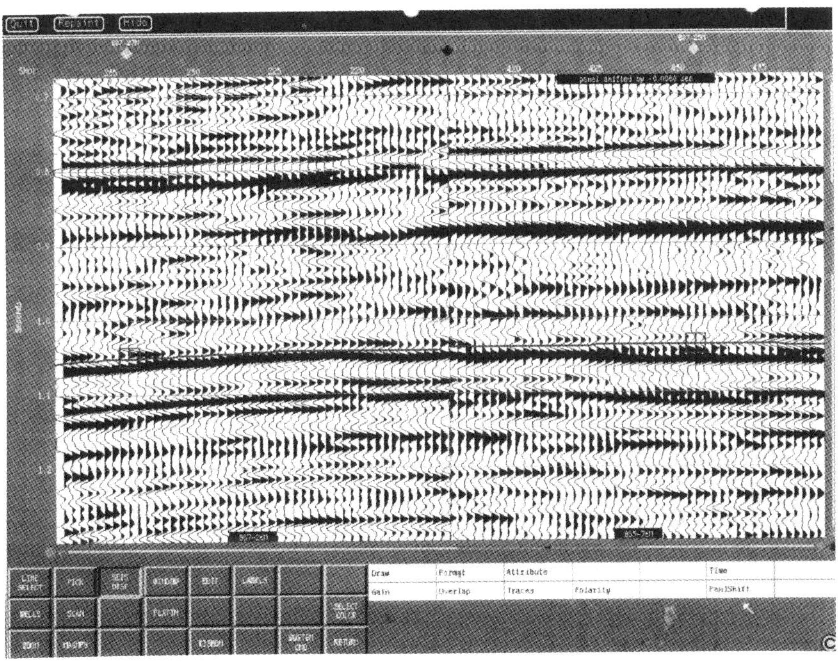

Sierra Geophysics, Inc.

Fig. 1–9 Screen dump

the part of the screen image you want included, give it a name, and give a command to save it. Then it is just a saved image. You can't interpret on it, or otherwise change it, except to delete it. Saved, it is available to make a print from. That print doesn't include commands or other peripheral information (unless you chose to include them). It is the kind of print you would use for formal presentations.

Another form of hardcopy is photographic slides. There are two ways to get them from an interactive interpretation system. The simplest is to just take a picture of the screen. Outside light can be blocked by a kind of cone, but with rectangular cross section. The large end fits against the computer screen, and a 35 mm or other camera is placed at the small end, so a picture can be taken. This works best if the glass of the screen is flat, as some are. But most screens, like ordinary television screens, are curved enough to distort straight lines a bit.

Another way to make slides is with a special camera made for photographs from computer images. It doesn't look like a camera, but is a large box connected to the system by wire. It has a flat screen inside it, that can display whatever is on one of the screens of the workstation. A conventional camera is fixed rigidly at an exact distance from the screen. You give a command, an image is displayed on its screen, and the camera takes a picture of the image. With everything held rigidly in place, a long exposure can be made without problems caused by motion. The device isn't fast, as it takes a while to expose each frame. But you can instruct the system to have it photograph a group of slides one after another automatically. You can set it to make a number of slides at night, for instance, when the computer might otherwise be idle.

THIS BOOK IS ABOUT OIL AND GAS

You won't find a lot of references to oil and gas in this book. But THAT IS WHAT IT'S ALL ABOUT. Loading data, picking horizons, making different kinds of display, are not ends in themselves. You will normally be doing those activities to find oil and gas. You might sometimes be evaluating a lease that has poor prospects, expecting to let it expire. Or you might interpret to determine where to locate a water-injection well. So you won't always be directly looking for oil and gas. But even these other reasons are indirect ways of getting more oil and gas.

The point I'm making is that an elegant system, a fancy display, a neat calculation, or a smooth operation of the system is worth nothing in itself. You are doing all this computer stuff to FIND AND DEVELOP OIL AND GAS FIELDS.

2

Interpretation Start

When starting an interpretation there are several aspects of the area to find out about. If it is to be your first interpretation on an interactive system (Fig. 2–1), you need to know some of the functions the system performs and how to give the commands for those functions. Later you can learn more, as you need additional functions.

Whether you are doing the interpretation for exploration or field development, the basic functions of the system are the same. Many of the functions are also the same whether you are interpreting 2-D or 3-D.

The similarities make many of the procedures for interpreting 2-D and 3-D similar or the same. The commands for the two are given in the same way. The system can keep track of the reflection times of the picks in the same way. Etc.

Because of these similarities in requirements and procedures, most of the topics in this chapter and in the next two chapters apply more or less the same to either 2-D or 3-D data.

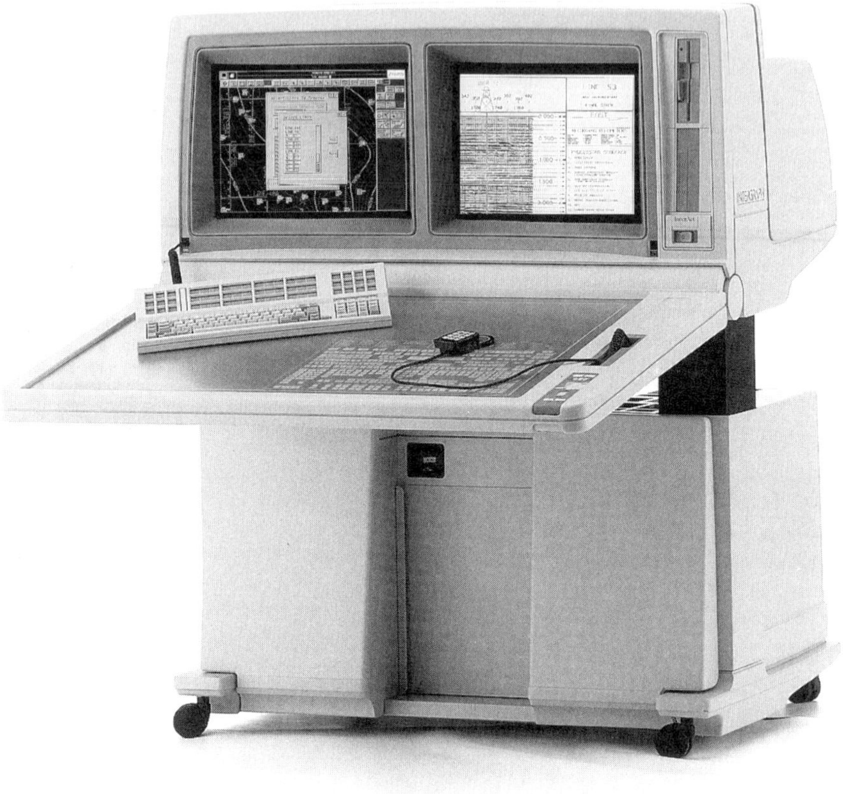

Intergraph Corporation

Fig. 2–1 Workstation

FIRST THINGS FIRST

There are some general things you need to consider at the start of the interpretation of an area.

Find out the purpose of the interpretation you are to make. You were assigned the area to interpret for some reason, maybe to select a location for a well, reconsider a fault interpretation, look for new prospects, or just add to a regional picture.

Find out the time schedule for the interpretation. There may be a deadline because a lease sale is coming up, because a rig is waiting on expensive standby time, or because a lease is about to expire. There may be some urgency because your time will be needed later for another project.

Or the need may be for a slower, more thorough investigation to get all the information you can about the area.

There isn't any set amount of time that an interpretation should take. I have occasionally made what I call a "five-minute interpretation" by looking at a couple of sections and deciding that there probably is a structure on them, or that the data is poor, or something. The value of that kind of "interpretation" could be to decide whether to spend the time to look at a farmout offer from another company, or whether it is worthwhile to make a new interpretation of the area. And I once made a five-year interpretation, working a large complex area, updating as more data came in from new shooting and from purchases of old data. Both the five-minute and the five-year interpretations were useful in finding oil.

To work on an expensive interactive interpretation system that is probably in much demand, you need to know when it will be available to you. If your interpretation is one that is vital to the company, you may have a high priority on the system. But if it is just one interpretation among others, then you may have to find times when you can get on the machine. It may be that you can work most consistently by using it after normal working hours. If you are a consultant, paying a fee for use of the equipment, it may be cheaper on weekends and at other odd times.

A seismic interpretation isn't a test of what you already know like you encounter in school. To make the best interpretation, you need all the information you can get, not just what you already know. Get whatever geological background information you can. Read articles on the area. Quiz your company's geologists about the regional geology, the kinds of prospects expected, the formations that are known to produce or expected to. When a geologist starts talking about your area, pay attention.

Get out any old seismic reports on the area and on nearby areas and similar ones. Read them carefully. The writers have been over many of the problems you are about to encounter. You can profit from knowing what they discovered and what puzzled them. You will make your own interpretation, and may not agree with them, but you can make it quicker and better if you know the problems other people had and the ideas they came up with. Or you may want to look the area over first, to get your own ideas about it. That's good, it's a scientific approach. But then be sure to also learn the other interpreter's ideas. That too is scientific, trying to not miss any appropriate thinking.

Find out what people want from the interpretation: maps of certain specific horizons and time interval maps between some horizons, or just the best idea you can get of the area. Your boss may tell you what to map, or you may find out in a talk with a geologist what horizons would be most useful. Sometimes there will be an assignment or even a contract (if you are consulting) laying out exactly what horizons are to be mapped.

In other situations you may be expected to do the best you can, maybe mapping one or two horizons in a poor-reflection area, or a great many horizons in an area that has good enough data for automatic picking to speed the interpretation.

The way the area was shot and the way it was processed are things you need to be aware of. If it is on land, you may find that multiples form a pattern the reverse of that of the hills and valleys. An offshore area may have its poor reflections where the seabed is soft, or may have its reflections overwhelmed by reverberations where the bottom is hard.

If you have the time and look up all this information (the same information you'd need for working the area on paper), you will become familiar with the area before being confronted with a machine to do your interpretation on. You can get the data loaded in the system and be ready to go.

LOAD THE DATA

Before interactive interpretation can begin, the seismic data for the area must have been shot and processed, and then must be loaded onto the system so any part of it can be called up instantly. The loading consists of running the data processor's final tapes and having the computer copy them onto its built-in hard disk or a removable optical disk.

The disks are considered external memory (or storage). When data from them is to be used, the computer copies that information into its temporary internal memory.

The kind of optical disk used in seismic interpretation is WORM type (WORM for Write Once, Read Many times). Information is stored on the disk, not as magnetized spots, but as tiny pits in the disk, formed by the heat of a laser and read by a weaker laser. Information can be put on the disk, but there is no way to erase anything from it. After the data is loaded onto a blank disk, it is on the disk permanently. This situation is reassuring. No wrong command can destroy the seismic data on the disk, that is, the traces you are going to interpret. But the picks you make in interpreting will be stored as a magnetic code on a hard disk, so they are erasable. That too is reassuring. You wouldn't want to preserve picks you are just trying out to see how they work.

Data on a hard disk is, like other data in magnetic form, erasable. When you are working on an area, software keeps you from deleting the basic data. For instance, the interpreting commands don't include a command that deletes that mass of data. The hard disks used in seismic interpretation are permanently installed, not removable like optical disks.

There are reasons to delete basic data, though. In systems with storage on a stack of hard disks, the capacity may be enormous. But even so, as more areas are loaded on hard disk, the disk or stack of disks starts getting full. Some of the areas must be erased to make room for other areas. The permanent storage of the data is on the processor's tape, from which the data was loaded on disk. The tapes are normally kept in a safe place with other tapes. All your interactive interpretation will use the copy on disk.

In some systems, to speed up access to the quantity of 3-D data, the data may be put on disk not just once, but as two or three versions. In one version there is a copy of every sample of every trace in every line that was shot in the area. That is all the seismic data there is in the project, so any group of samples, for instance those that make up a crossline or a time slice, could be selected from it and displayed. But that kind of selecting takes computer time, and in some systems might make the process too slow to be practical. So, in addition to storing all the inlines, the data may be sorted into crosslines and stored in that form so any crossline can be called up quickly. And the data can also be stored in the form of time slices, so they can be called up quickly. That makes three complete sets of the same data in different arrangements on the system's hard disk or optical disk.

Some systems use those varied arrangements, although it takes three times the storage space. Other systems depend on greater computer power to speed the arranging of samples, so they can have just one set of data on disk, and call up other sets quickly enough without needing to have them already sorted out on disk.

There is a balance between computer power and mass storage (by computer power I mean the ability to quickly perform complicated operations on data). To take an extreme and unreal example, if you had a computer that wasn't very powerful and disks that held so much data that you could consider their capacity endless, then it would be practical to store many copies of each line. You could store them all in wiggle-trace form, another copy in variable density form, copies in different attributes, etc. Then to display any of them on the screen you would just call them up. Or if instead you had a tremendously powerful computer without much disk storage, you would store each data sample only once, and let the computer calculate each form "on the fly", that is, only when you ask for it.

Of course you don't have either extreme. In general, if you have a powerful computer you also have considerable disk storage, and if you have a lot of storage room on disk, you have a powerful computer. So the system uses some appropriate balance between storage and generation as needed. If the computer is powerful enough, the system may form the displays as they are needed, even though it does have room for a lot of data.

Data isn't always obtained directly from tapes. It may be received by wire from another computer in the company's local area network (LAN) or even by telephone from some far-off place. Or the data may be obtained by scanning paper sections. Once the seismic data is loaded, you are through with the tapes or other source you got the seismic data from.

There are some steps that have to be taken at the start of any computer project. They are like the ones I encountered with my first computer, which was bought for word processing. I discovered that there were several things to be done before I could start using the computer for writing. The program had to be copied onto another disk, the disks that were to hold my data had to be formatted, and the word-processing program had to be "initialized", that is, told what hardware it was to work with. All that was far too much for a novice to handle. I didn't even know what the words meant. The salesman did those things for me.

Loading seismic data is like that—a task for a person who knows the procedures, not for someone who is just starting. You need a specialist or an old hand to do that for you. Even if you learn enough to load data yourself, it may be better to have an expert do it, so you can concentrate on interpretation. All these routine tasks that must be done exactly right, and that are preliminary to the real work on the system, are tasks that may one day be accomplished automatically by the computer. It will be like the cameras that sense what kind of film they are loaded with, and then set the automatic exposure control to the speed of that film.

In loading seismic data, there is a great difference between 2-D and 3-D. Since 3-D is in uniformly spaced lines and traces, it is easier to load than 2-D. For 3-D you give the system a starting point, direction of the first line from that point, trace interval, and a few more simple parameters like those, and leave the system to do the loading. But 2-D lines are separate, often not even parallel, maybe with bends. At least some of the parameters for each line have to be given individually. The information may have to be dug out of reports and maps for each line. In a way, loading a 2-D line is like loading a whole 3-D area.

In addition to the seismic data, there is other information to be loaded on the disk. Some of that information may be in a database.

DATABASES

Databases are necessary to interactive seismic interpretation. A database sorts and keeps track of information. To interpret an area, in addition to the seismic data, you need a lot of peripheral information, especially of the type that is normally presented on maps (shot point locations of the current and older projects, terrain features, land ownership and leases, well locations, etc.). That information comes from a variety of sources, and must be brought together so you can use any part of it at any time. Assembling all the information is a major task for a company to undertake, especially if it has a large backlog on paper.

First, a coordinate system and map projection must be chosen. Then all the existing maps must be converted to that form. That alone can occupy surveying experts for a long time. Compromises have to be made where different maps don't agree. The result isn't perfect, but you get the mapping in as good shape as you can, and then pretty much accept it as correct. Then any new data can be gathered with an effort to make it compatible with the database. That is, the existence of the database more or less requires that the company have a policy that future surveying use the same coordinate system. Unfortunately, this leads to rigidity as opposed to innovation, but it simplifies blending future data with the old.

The different kinds of data are put into an X-Y coordinate system. It is easier to work with than latitude and longitude because distances in X-Y systems are decimal, in 10s and 100s, instead of the 60s of degrees, minutes, and seconds of latitude and longitude. But for a system that is to cover large areas, latitude-longitude may be preferable. It fits any part of the earth, large or small. So databases are often kept in that form, and then converted to X-Y coordinates for use. The conversion is done automatically by the computer.

Locations of wells are put in the coordinate system you are using. Logs are digitized and added to the database. Provision is made for a geologist's picks of formation tops to be added or modified at any time. Land descriptions in the legal form of township and range are converted to the database standard. You can call up information from a map on the screen, instead of from a list, getting answers to questions about wells, leases, and seismic lines (Fig. 2–2).

Your company may not have completed, or even undertaken, the construction of a database. Or maybe it was started and then came apart, so people gave up on it. If your company is farther along in computer use, it may have a complete database, that it spent much time getting to fit the company's needs. Then it will probably insist that all additions to it be in a form that is compatible with it, or that can easily be made so.

There can be great difficulty making different sets of data fit together into one database. Each has its own peculiarities and may have its own errors. So building a database from the company's accumulation of data is a long, difficult, tedious chore. Some old data may be so hard to fit in that it is best left out of the database.

Much of the data is loaded into the computer in "batch" mode, that is, with the computer working on it "in the background", not apparent on the screens. Batch mode work is often saved for nighttime, so it can go on without bothering people.

Making a database of location information is such a complex task that there are companies that specialize in putting databases together. They may combine government maps with surveyors' measurements, then check

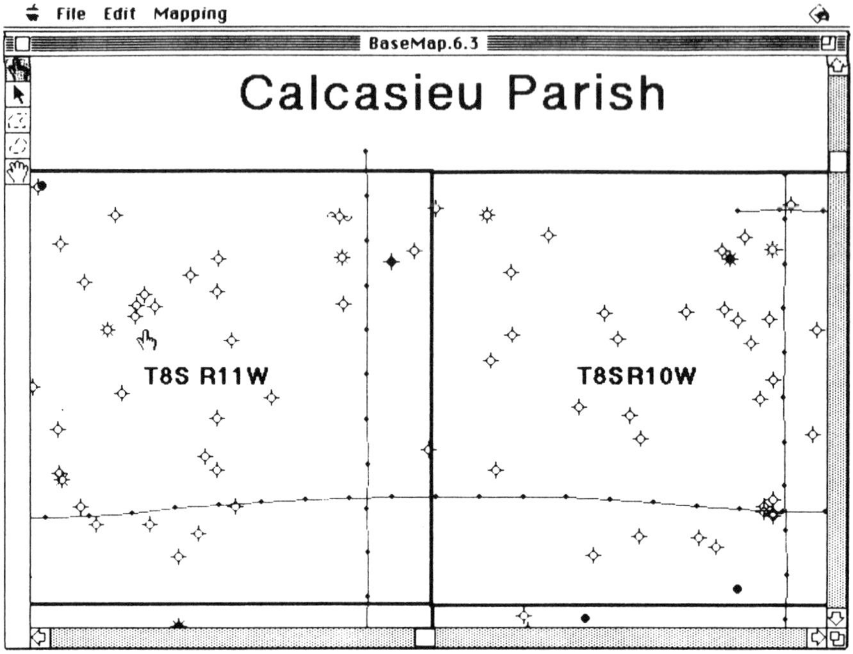

a Base map for selection

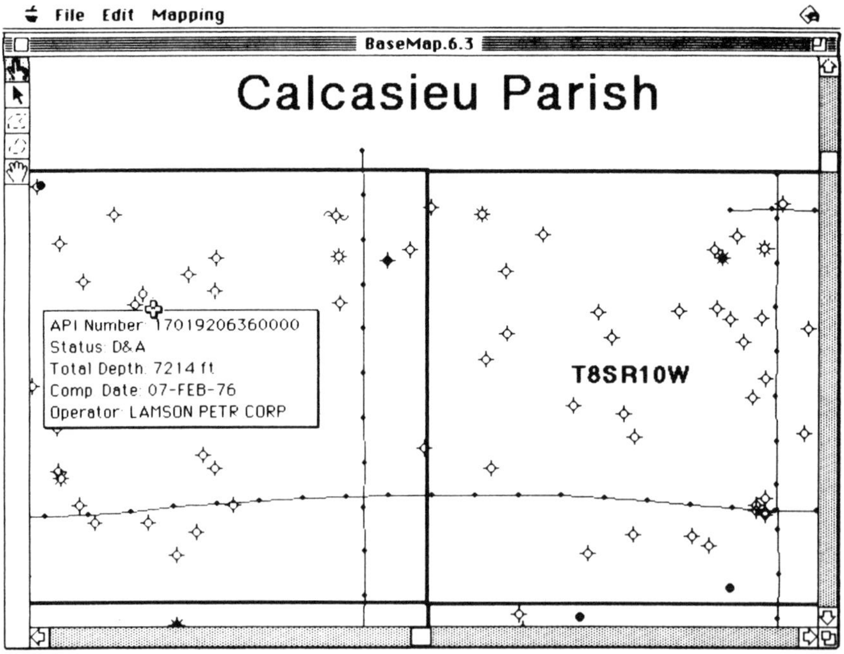

b Well data selected

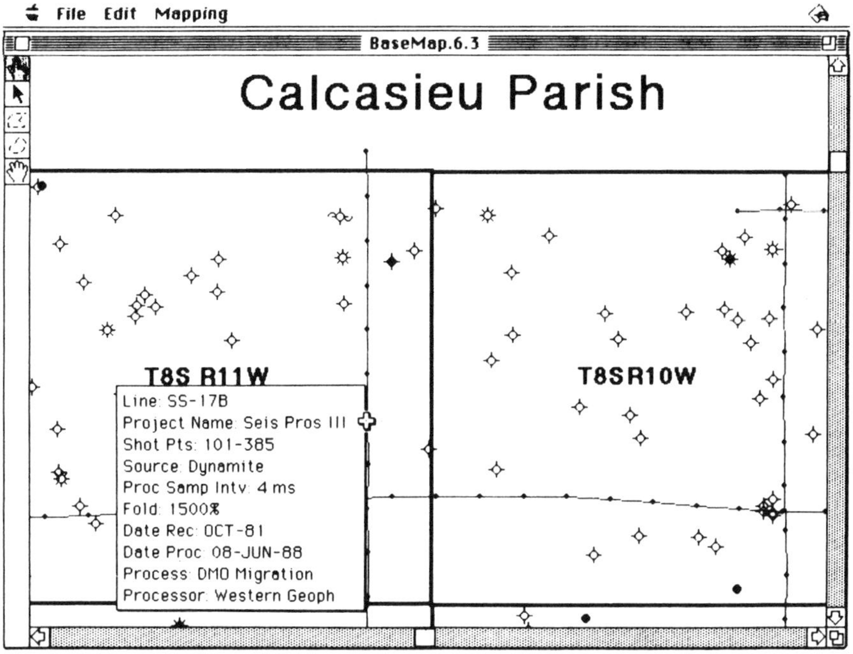

c Seismic line data selected *GeoQuest International*

Fig. 2–2 Calling up information from database

locations of wells and other features against air or satellite photos. Information about wells, leases, seismic projects, and the like can be called up and read or displayed on maps as needed.

FILES

Data in computers is organized in files. In this sense a file is a collection of information that can be called up together, like a file of papers kept together in one folder. When you pull the folder out of a cabinet, you have the whole file.

In interactive interpretation, the picks for each horizon may be kept as a separate file. Then you can call them up and modify them conveniently. When you want to work on another horizon, you call it up. A change on one horizon does not affect any others. You can display as many horizons as you want at the same time, but only work on the one that is active.

If you delete one file, that doesn't affect any others. So the harm caused by an accidental deletion is limited to the one file.

Although files are thought of as the units of computer storage, databases and interactive interpretation are complex. They may use the word "file" in confusing ways, so there can be files within files. The data for an area may be considered a file, the interpretation another file. But within the interpretation file, the individual horizons are stored so they can be called up separately, so they can be considered files. With computer use developing rapidly, words are bound to be used in confusing and even conflicting ways. In this case, just think of a file as any unit of information that you can call up with a single command, whether or not it is part of a larger unit or contains smaller units in it.

LAYERING

The word "layering" has recently come to mean wearing several layers of clothes. But in databases and displays from them it means having different aspects of a display stored separately in the computer so they can be combined like transparent sheets for an overhead projector.

The geographic coordinates can be in one file, the streams and coastlines in another, with others having leases, wells, shot points, horizon times, contours. Then by calling up the appropriate files you can show the landman seismic contours with leases, or you can look at shot points, reflection times, and contours together (Fig. 2–3).

That has the tremendous convenience of computer handling of the layers, as opposed to any physical overlay method. You can combine as many layers as you wish without getting a dingy background like you would from many layers of film. The layers are always in perfect register. You can easily try different combinations. You can make a printout of any combination.

COMMANDS

There are two basic kinds of commands. One kind causes something to happen at the time, a "do it now" command. The other kind of command sets a control that will influence future commands, a "do things this way until I tell you to change" command.

The first kind is a one-shot deal; the second has a continuing effect. An example of the first type is a command to call up a certain section onto the screen. As soon as you give the command, the section appears. A command of the second type might establish a color arrangement for the section. If the command is for the peaks to be blue and the troughs red, then, until it is changed, all the sections you call up will have blue peaks and red troughs.

You will use some of the do-it-now commands regularly, much more often than others. These commands will be the ones you spend your time with, the main ones that allow you to do interactive interpretation. Of them, you will most often use the commands for picking horizons. There are several of those, for picking by hand or with various degrees of automation

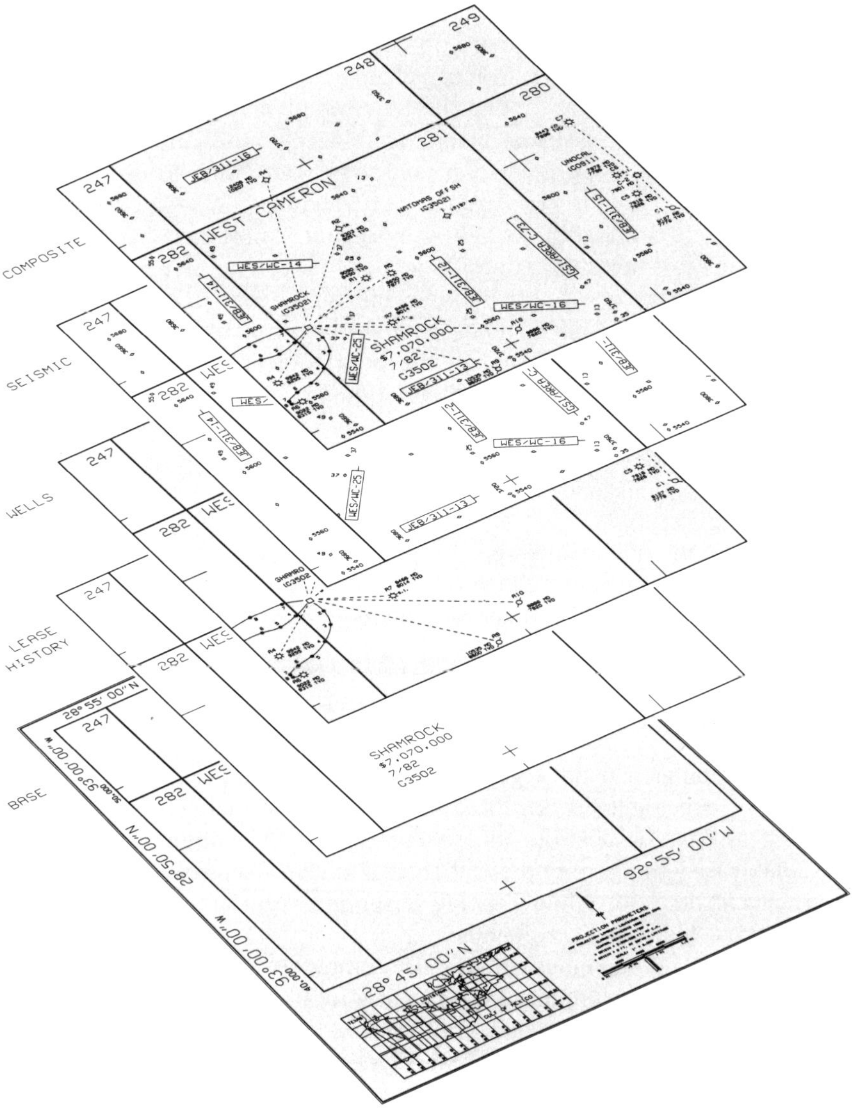

Energy Graphics

Fig. 2–3 Different layers of data

in the picking. Other commands you will use a lot are those that call up sections and slices so you have something to do your picking on. Another is the command to update the map on a screen. That one you will use fairly frequently, unless your system is a type that updates the map automatically as you pick.

Some of the continuing commands may already be set in the system as defaults, standard settings that you can use if you don't want to go to the trouble of making choices. You can change from any default to a different setting whenever you want. Then the new setting stays until you end the session. And in most systems, when someone else uses the system your settings are not in effect, but when you enter your name to use it again, your settings are back, as though they were defaults.

While you are learning a system, you don't know many commands, so you have the computer perform functions that you know the commands for. You wind up getting things done, but they aren't exactly what you want done. This isn't too different from working on paper, though. With paper, you are similarly limited. If the print shop is closed for a while, you might have to get by without new prints of sections that you want. Or you run out of red pencils; then until you get some, you can't mark any horizons in red, but have to get around it with orange or something, or maybe with plain black-lead pencil that you plan to erase later and go over with red.

The commands in interactive interpretation are given with the help of input devices and menus. The input devices are mechanisms that transmit your commands to the computer and the menus are the lists on the screen that you choose commands from.

WAYS TO GIVE COMMANDS

A computer can do a great many things for you. For you to tell it which of those things to do, there must be a great many commands. To make it as easy as possible to use this power, the commands must be reasonably easy to learn, to remember, and to give. To resolve the problem of having it do a lot while keeping it simple, there are various devices and methods for giving commands.

In many non-geophysical uses of computers, commands are given just by keyboard. A computer keyboard is a lot like a typewriter keyboard, with the letter and number keys in the standard "QWERTY" arrangement, but with a number of extra keys that a typewriter doesn't have. There are special keys to give certain commands, and at one end of the keyboard there is a "numeric keypad", that is, a second set of number keys, but arranged in columns and rows like the keys on a calculator.

The normal characters on a computer keyboard can work either like those on a typewriter, so you can write text with them, or in a different mode to give commands. The different mode is sometimes achieved by pressing the key labeled "Control" (or CTRL) at the same time as another key. That makes the other key a command, in the same way that the Shift key with another key makes that other key a capital letter. At other times, when the computer expects to receive a command, you don't need to hit the control key.

With most interactive seismic systems, you will use the keyboard just at the start of a session, but some systems give you the option of using it to give later commands. You may type a few words: your name, department, the name of the seismic project, etc. Typing on a keyboard is of course an easy way to convey words to a computer. Any time you use the keyboard to give an instruction that is made of a number of characters, you will hit the Enter key (or it may be labeled Return) to let the computer know you've finished typing characters, and that it is time for it to do whatever the characters call for.

Several of the ways of giving commands involve moving a cursor around on the screen with one device or another. The cursor is an image on the screen that indicates a specific point. It may point out the latest position at which you gave the computer an instruction, or you may move it to select some choice from the program. Or you can draw lines with it, for instance by moving it along a reflection to mark your picks.

When you move the cursor to something you want the computer to do, an equivalent of the Enter key is needed to tell the computer you've got there. This equivalent is usually a button on the device that moves the cursor. As you move the cursor around on the screen, nothing happens. You move it in and out of "boxes" (outlines around commands) and nothing happens. But when you press the button while the cursor is inside one of the boxes, that signals that you are at your destination, and the computer obeys the command in that box.

There are several different devices for moving a cursor and giving commands (Fig. 2–4). People keep thinking up gadgets that make commands easy to give.

- On the keyboard there are four keys, labeled with arrows, that move the cursor in the four directions. The keyboard may be on the desk or in a drawer so it can be out of the way when not needed. All the systems use a keyboard; almost all systems also use at least one other device for moving the cursor and otherwise giving commands.
- A mouse is a flattish object with a ball protruding downward from its under side. Moving the mouse around on a desktop or other surface rolls the ball, and the rolling is sensed and transmitted by wire to the computer, where the cursor moves in imitation of the motion of the mouse. The mouse has from one to three buttons on it to give commands to the computer. It may have been first called a mouse because buttons are positioned like a mouse's eyes or ears, and the wire trails behind like a tail. Pressing a certain one of the buttons is the usual way to indicate that you want something done. Since the button makes a clicking sound

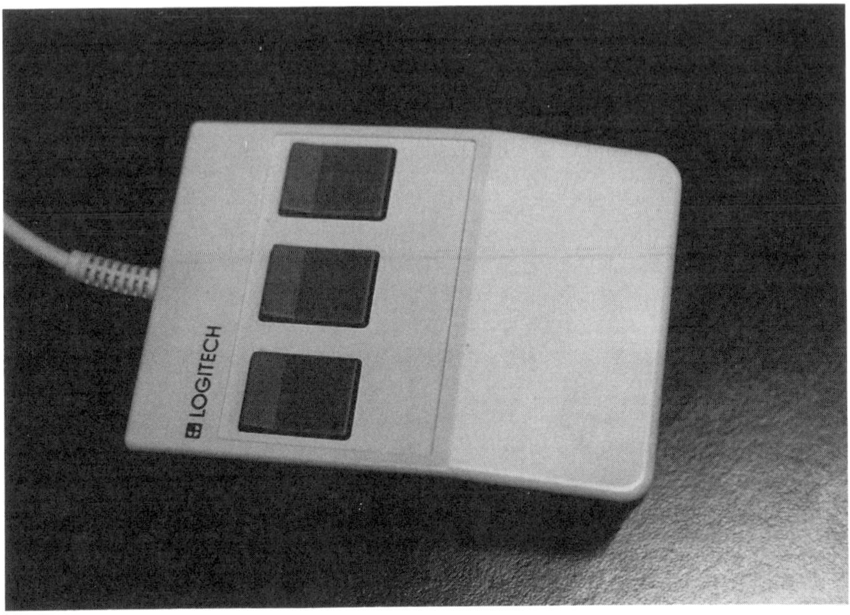

a Mouse

b Trackball

c Puck

d Keyboard

Fig 2–4 Cursor movers

when depressed, pressing it is also called "clicking". Move the mouse around, get it where you want it, and click.

- A trackball is like a mouse lying on its back. That is, the ball is set in its upper surface. Instead of moving the whole thing around, you rotate the ball with your fingers. This device too usually has buttons.
- A puck is like a mouse, but without the ball, and is restricted in where you can move it. The puck is moved over a graphics pad, also called a graphics tablet. It usually has more buttons than a mouse does. A typical puck may have buttons for the numbers 0 through 9, and the letters A through F. A puck may have got its name from being slid around the desktop like a hockey puck on ice. A computer puck is sometimes called a cursor. That's confusing, but it's easy to see how that use of the word came about. You move the puck (on the graphics pad) in order to move the cursor (on the screen). What is an onlooker to think you mean, seeing your hand move, when you say you are moving the cursor? To avoid this confusion, I'll use "puck" for the instrument and "cursor" just for the image on the screen.
- The buttons on a puck are used for special commands that are used a lot. One button may give a command that is also available on the screen, like enlarging part of a section, but that is used so frequently that it is handy to have it at your fingertip. One button may indicate that you want to start a segment of picking, and another that you are at the end of that segment. A mouse, with fewer buttons, may use the keyboard for some of those special commands. The keyboard isn't already in your hand, but it has the advantage of having the whole alphabet, so the buttons can be chosen to be reminders of the functions they call up.
- The puck has crosshairs in a transparent window, to position precisely over some point on the pad. In non-seismic uses of computers, a graphics pad may have symbols on it that are pointed to by the crosshairs. A seismic interpretation system may use the symbols too, but in most seismic systems the crosshairs are used just to enter data from paper sections and maps.
- The graphics pad is a surface with an internal network of wires to sense the position of the puck. The graphics pad may be a flat board lying loose on the desk, connected by wire to the computer. Or it may be built into the top of the desk, so it isn't visible. The location of the pad may not even be marked, so you can find it only by moving the puck around until you see the cursor dart across the screen or an indicator light on the puck come on. When those things happen, the crosshairs of the puck are in the area of the graphics pad.
- Some manufacturers don't like to have an outline on the desktop to show where the edges of the imbedded graphics pad are, so users often put

their own outlines on the desktop, maybe with tape. A system with its graphics pad hidden is like a portable computer without a handle. It can be made to work, but an extremely simple addition makes it much better. If it's not marked, you at least rather quickly become familiar with where the graphics pad is on the desk.

- A mouse too may allow you to get lost. The mouse has no specific area that it has to be on. You can move it anywhere on a desk. If you move it too far, the cursor goes off the screen, and you have to move the mouse around until the cursor reappears on the screen. When you have a section enlarged so the screen is filled with a magnified version of just a small part of the section, the problem is worse. Smaller movements of the mouse tend to send the cursor flying off the screen. The problem when it is off screen is that you may not know which way to move the mouse to bring it back.
- A stylus operates much like a puck. The stylus looks like a pen with a wire extending from its top. You hold it like you would a pen, and move it over the surface of a (maybe hidden) graphics pad, and the cursor moves in response. The stylus may have a button on the side, to give commands with. Or the function of the button may be performed by a spring-tip, so you just press down with the stylus.
- There are other devices that you can use to give commands to a computer, either to move a cursor or to make some other choice. A wheel may be set either horizontally or vertically, so it can be spun to make some choice or other. A separate keypad, like a numeric keypad, may have its keys assigned to special commands like the keys on a puck.

A cursor on the screen can be any shape. For interactive interpretation, it is some shape that stands out well and points to a place on a display, or to a choice you make from a list. The cursor used by a system may not stay the same shape. When you move it from one part of the screen to another, it may change abruptly from one shape to another as it crosses some boundary. And when you call up a different display on the screen, the cursor on the new display may be a different shape from the one on the earlier display. A cursor that is useful for making picks on a section may not be the best for selecting a command from a list.

The cursor may be an arrow. If it is, it indicates a spot at the tip of the arrow. It may be a plus sign, with the point at the crossing of the two lines. A particularly useful cursor is a plus sign with variable-length arms, so you can choose for them to be short so as not to get in the way, or longer, up to so long that they reach to the edges of the screen. There are several things this long-armed cursor can do for you. Its horizontal line acts as a convenient sort of t-square to let you see whether a point you are picking is higher or lower than some other point, maybe on the other side of a fault. It may make gentle dip evident by contrast, dip that you wouldn't otherwise notice. The vertical line can show when the cursor's center is directly below a place where a fault cuts a horizon, or whether a sag in a reflection is directly below a feature that may have low velocity indicating that the sag is a velocity pull-down.

With displays on two screens, cursors can tell you things about the relationships between them. As you move a cursor about on a section on one screen, a cursor on the other screen can be automatically moving to show where those positions are on a map. For instance, if you have a section of Line 18 displayed, with the cursor at Shot Point 632, the cursor on the other screen will be at that point on the map, Shot Point 632 of Line 18. As you move the cursor along the section, the other cursor runs along that line on the map. If you have parts of two sections joined at their intersection, and move the cursor along that display, the one on the map will run along to the intersection point, then make a sharp turn and run along the other line. In another system the cursor will be on just one screen, and will jump to the other when you move it beyond the edge of the first screen toward the second. In that situation a mouse's lost cursor could show up on either screen.

For choosing an item from a list, you may highlight a command's box. That is, one of the choices may be highlighted by having its background in a different color. That makes it easy to see which choice you have made, without your having to position the cursor at some exact point. As you move the point across the border between items, the special background color vanishes from one label and appears on the next. The cursor may still be visible pointing to the command, or may disappear so you only see the changing highlighting.

A system may use pictures to represent some of the choices. Pictures are easy to recognize and to remember. They are also understandable across language barriers. So "icons", simple little sketches, are put on computer screens to represent commands, and the user can select a command by moving the cursor onto the appropriate icon and clicking (Fig. 2–5).

Considerable ingenuity has gone into devising icons that people will easily understand to mean certain actions. One function that lends itself easily to pictorial description is that of selecting the parts of a wiggle-trace section to be filled in to make a wiggle/variable-area section. It's easy to point to a picture of a wiggle with no fill-in or one with only the peaks filled in. And having it in the form you are expecting to see on the section helps you avoid getting the wrong one by mistake. In that particular kind of choice, even the most skilled with the program may get mixed up if the choices are given in words, but when examples of the various forms of display are there to be pointed to, people will rarely make mistakes.

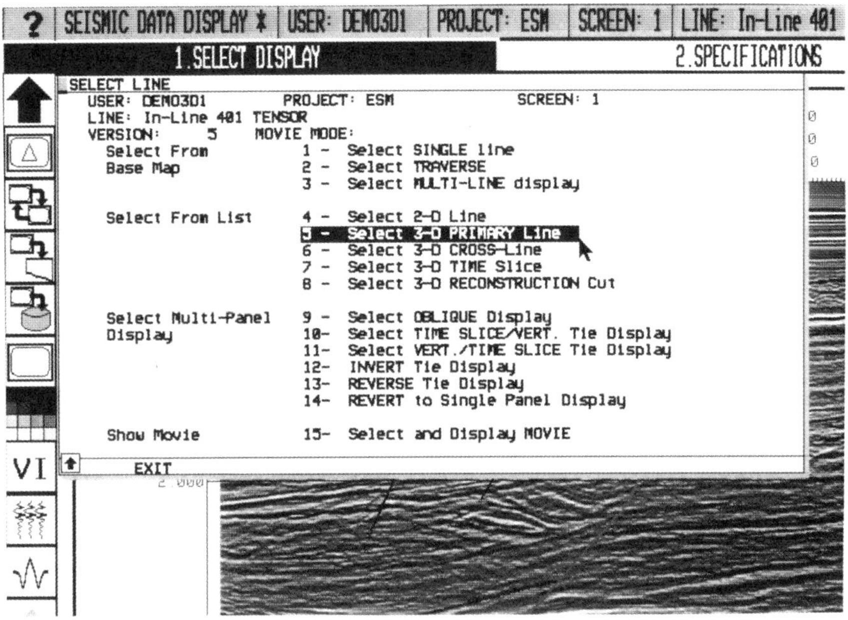

GeoQuest Systems, Inc.

Fig. 2–5 Highlighting and icons

Some commands lend themselves to pictorial representation in a straightforward way, and others do not, so the systems that use pictures use them for only certain functions. Other functions are expressed better in words. It would be difficult, for instance, to design an image that people would automatically understand to mean "wiggle with frequency overlay". So the commands on the screen may be made up entirely of words, like "Interpret", "Choose Colors", or "Update Map". Or some items may be expressed in words and others in pictures. The trick is to make the act of choosing as easy and natural-feeling as possible.

A "scroll bar" is a sort of long icon that is used to show you what part of a section you have on the screen, and to change the display to some other part. There are usually two scroll bars, a vertical one along one side of the screen and a horizontal one along the bottom of the screen, the two representing the vertical and horizontal extents of the section (or other display). The whole bar represents the entire height or length of the section, and a part of it in a different color or shading shows what part is on the screen. To change the display to a different part, put the cursor on the appropriate bar, at the place where you want the new center of the display to be, and click. The display vanishes and the new part of the section is drawn in its place (Fig. 2–6).

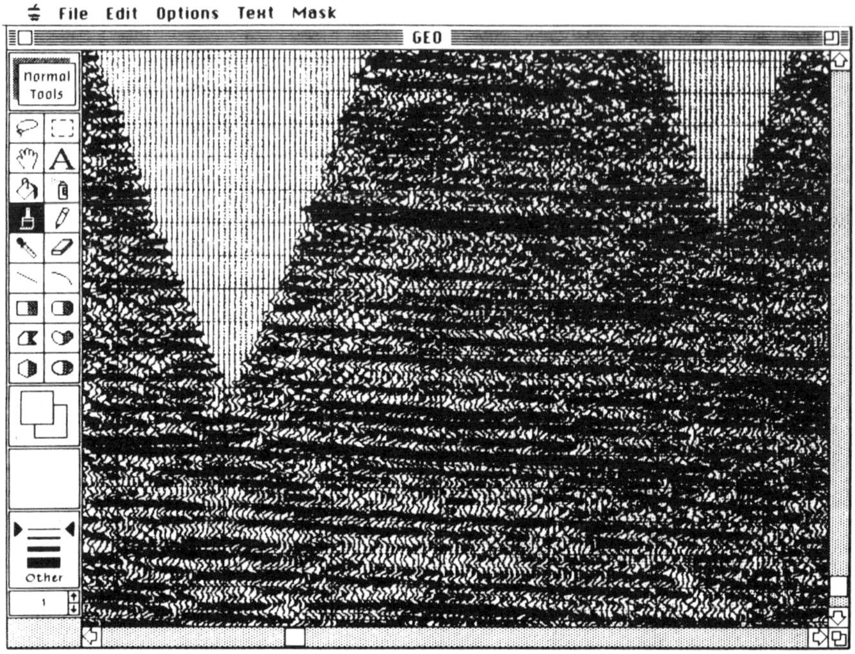

GeoQuest International

Fig. 2–6 Scroll bars

Another way of seeing where you are on the screen is with a column of numbers, all but one invisible. As you move up or down, different numbers become visible while others vanish. This is useful to tell the computer what part of the vertical extent of a section to display, by pointing to what you want to be the top and bottom reflection times on the screen. This is a nice, intuitive way to select times. If you want to have the times on the screen start at about one second, then, since the exact point doesn't matter much, you can click when a number somewhere near 1.000 appears. That is closer than you select with your eyes when you choose to look at part of a paper section. But if you want it to start at exactly 1.000, maybe to make a hardcopy for a report, you may have to move the cursor up and down a bit, hunting for just the right point. That can be difficult.

Notice that the becoming-visible technique is used for something predictable like a series of numbers. You know which way to go to get a larger or smaller number. This system wouldn't work as well for selecting a command from a list, because you wouldn't be likely to know which way to go from one command to get to another. The other method, highlighting, works better for a list of commands.

One way to enter numbers is with a picture of a numeric keypad on the screen (Fig. 2–7). Then you just touch the numbers with the cursor, and click. So for numerical input, like a line number, you can either click on the line on a map, or go to the picture of a keypad and click on the digits that make up its line number. The picture may also include the arithmetical signs so you can use it as a calculator. It is fairly easy to use, pointing and clicking on the digits in sequence. And it keeps you from losing your place by turning away from the screen to a keyboard.

A cursor is excellent for keeping your place when you are interrupted. If the phone rings, or someone comes in to ask a question while you are picking a section, you may lose track of just where you were. But, if you don't bump the puck or whatever with your elbow, there is the cursor, patiently waiting where you left it. If the cursor is small, and you don't see it right away, just jiggle the device that controls it, and the cursor's motion will make it evident.

I LIKE A KEYBOARD TOO

Makers of interactive interpretation systems try hard to make the systems user-friendly. They don't want fear of computers to keep people from using the systems, and they don't want using systems to seem like typing. Many of them keep use of the keyboard to a minimum. To select a section to display, point to a seismic line on a map. To choose the range of times to show from that section, move a pointer to the times on a list. The idea is that it feels natural to people to indicate their wants by pointing to them. It does, and that is often a good way to select. Some functions, like picking

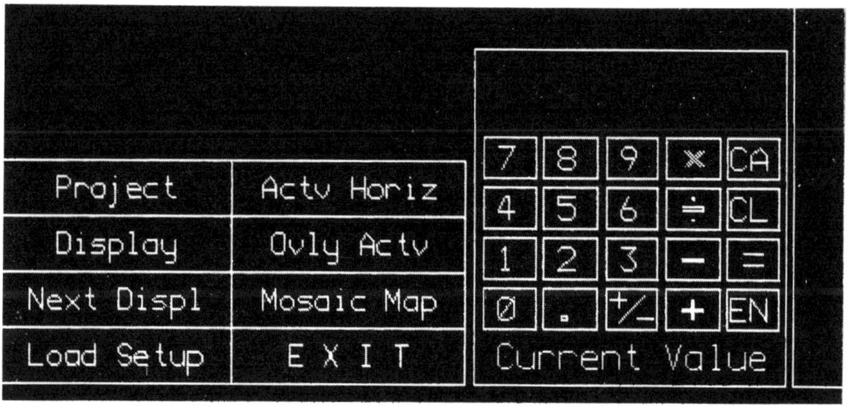

Western Geophysical Division, Western Atlas International, Inc.

Fig. 2–7 Numeric keypad on screen

and marking a piece of a section to be moved for correlation, are much easier with a mouse or other pointing device.

Punching instructions on a keyboard isn't as easy to learn as pointing on a screen. And typing out words is slower than pointing to words already spelled out. But that doesn't mean you will never want to use the keyboard. Typing number keys to give a line number or a seismic time is sometimes quicker and easier than feeling your way to where the line is on a map, or where the time is on a list, especially if the map or list is on a small scale, so the items are crowded together.

Picture yourself interpreting away, when your boss comes in. You'd found something exciting on Line 23, and had mentioned it to the boss. Now the boss is available and would like to see the interesting feature. In a bit of a hurry, you move the cursor around on a map, and press the button. Oops, that was Line 25! You and the boss wait while the system finishes displaying Line 25, and you start again. Feeling more rushed, this time you happen to hit an intersecting line. You feel inefficient; the boss is impatient. How much better if you could have just punched a 23 on the keyboard. A keyboard is sometimes a handy alternative to the other input devices.

COMMANDS WITHIN COMMANDS

To be able to use the many functions of a system, you need a list of commands so you know what functions are available, and how to ask for them. The list is long, and there isn't much room on the screens. Even if the screens did have room, it wouldn't be convenient to look through a long list every time you wanted to give a command.

Systems, as far as I know, now all use some form or other of a command "tree". To do something, you choose a general category of commands, then from that category, select a smaller category, and so on, until you get to the command itself. And to go from one specific command to another, you may need to go back a stage or so to some more general category, and then down to the new command. It is exactly like the chain of command in a military organization, government agency, or company. To communicate officially with a person of your rank or near it, but in another branch, you have to send word up the chain of command and let it work down to the person.

In most organizations, there are some unofficial shortcuts that keep things from bogging down completely. In interactive systems, there are also usually a few shortcuts that take you directly to some much-used commands. But much of what you do is still running up and down the chain of commands. Sometimes it is hard to find where a function is in the command structure.

The systems have the lists broken down into sets of different types of commands. Each list is called a menu, since it is a listing of items to choose from. A menu of general categories, the "main menu", is all that is on the screen at first. Then you can choose one of the categories from it by putting the cursor on it and clicking. This causes a menu of commands in that category to come on the screen (Fig. 2–8). You can click on one of its items, and maybe get another sub-list. At one stage or another, there will not be a sub-list, so the computer will perform the action called for. Systems generally have from three to five such levels of commands. That can include a great many instructions, a hundred or more. In most systems, when you get a sub-listing, the list it came from vanishes and the new list takes its place. So there also has to be a command to allow you to go back to an earlier, more general, list. Or the first list may stay on the screen when the sublist appears. Then, to move to the more general list, all you have to do is move the cursor over to it, and click.

Going down the tree is fairly natural. You just look for a category on the screen, click on it, and see a list of commands that are part of that category. Going upward, you don't see where you are going so well. There is a command to go up to the next larger category. It may be called up by pressing a certain button on a puck, or clicking on an icon of an arrow pointing upward, or clicking on a command in the form of words at the top of the present list. Whatever it is, get familiar with it. You will need to use it a lot. When you are in the early stages of learning to use the system, forgetting about it can give you a lost and helpless feeling, looking at a group of commands and not knowing how to get to some other command. Also, the commands are necessarily brief, so they don't take up much room and so you can read them at a glance. Some functions can't be described clearly in a word or two, so their meanings are not automatically clear to a newcomer to the system. You not only have to know where to find a command you want, but also have to remember its name.

Even with the commands organized in lists as they are, there can be a problem learning what functions are available, what the commands for doing them are called, and where they are on the lists. A system will have a manual in the conventional form of a book or a notebook. The manual can help you find your way through the list of commands. But the main problem is the same one you encounter in a store, in the library, in your office files. If you are looking for an article that you remember shows a good example of a northern Canadian salt dome, you have to decide whether it is filed under Arctic or Diapirs. It depends on where the person doing the classifying thought it should go. And that isn't necessarily where you might be likely to look for it. Similarly, is the command to flatten a section in a list of calculations, of forms of display, or of picking options? Or is it in a special list apart from the regular chain, intended to be handy?

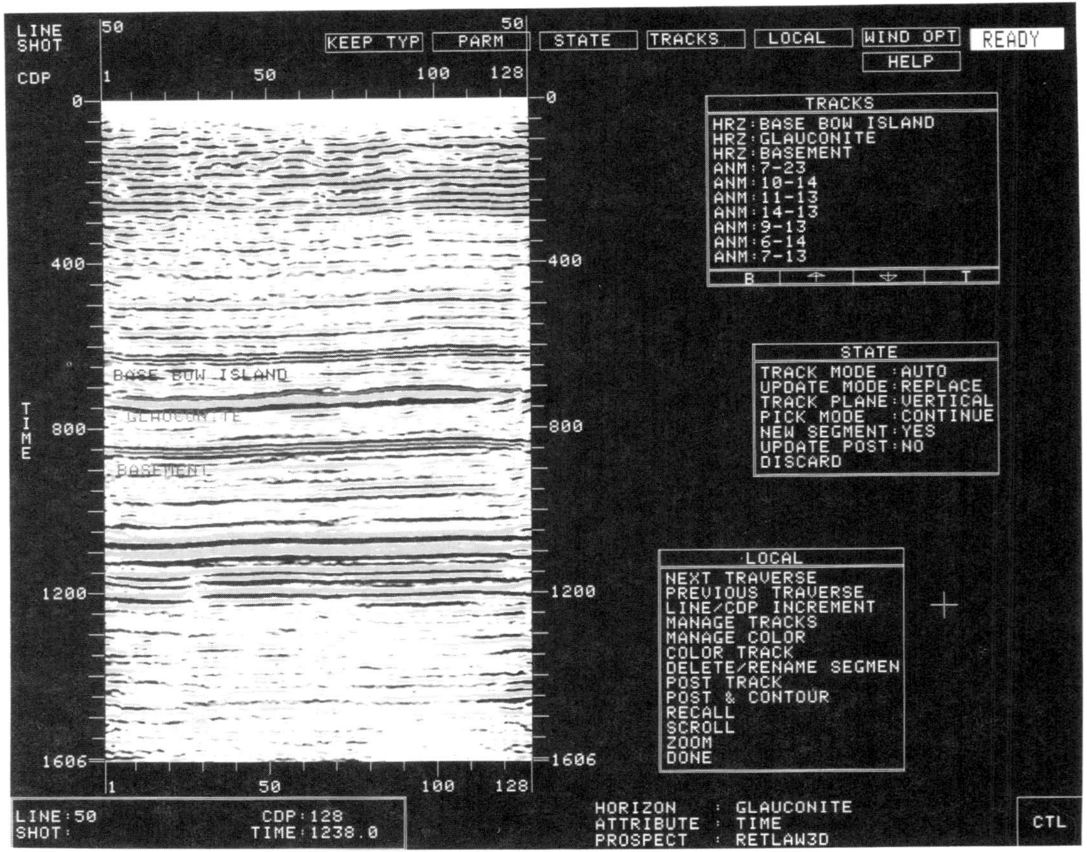

Sierra Geophysics, Inc.

Fig. 2–8 Menus in windows

In learning a system, you will have to do a lot of memorizing, so you can go quickly to any function you want. Just using the system will eventually fix in your memory the commands you use often. The main problem is before you get to that stage. Looking for the command to flatten can make you waste time running up and down the lists, especially if you happen to get a wrong idea stuck in your mind. Probably the best solution at this stage is to write a short list of commands you will use a lot or have had trouble finding. You can give an interpreting function like flattening by the name you tend to remember it by, and give the route through the lists to its name in the system. It may also be useful to make such a list of all the things the system does, in alphabetical order by interpreting function rather than by name of command.

So your own list, alphabetized by functions you want performed, each giving the steps through menus to get to that function, will help you around the problem. The very act of looking up and writing down will also help you remember some of the paths to functions. Don't worry about depending on the list instead of memorizing. The list is also a memory aid. Looking up the path to a function on the list will fix that path in your memory better than repeated searches by incorrect routes through the screen lists.

BOOT UP OR LOG ON

After being given some instruction in use of a system, and watching other people use it, you might be ready to do an interpretation on it for the first time. Naturally you will get lost in the system a few times, and have to ask for help.

"Boot" or "boot up" is a computer expression. All it means is to get the system started by turning on the computer and calling up the program you are going to run. Different systems may be booted in different ways. You may start the machine by turning a key in a lock, for security, like the key in your car. The interactive interpretation program may come up automatically, especially in a standalone system that is used only for interpretation.

If the program is on a larger computer that also runs programs for other people, there may be specific commands you give by keyboard to get the interactive interpretation program started. The computer is probably already booted, that is, it is running, doing all the other processes it is called on to do. It may be kept operating day and night. So to start interpreting, instead of booting, you "log on" to the computer, that is, request access to it, so it will work on your interpretation along with all the other things it is doing.

Getting started is usually performed by some short routine that you memorize and then go through each time like a ritual. It's different on different machines, in different companies, and maybe with different bosses, but once you start, it will be the same for you for some time. So don't feel bad if you get mixed up and do it wrong a few times at first. You will soon be so familiar with it that you do it automatically, like starting a particular car by depressing the gas pedal three times, turning on the ignition, and then gunning the engine a moment. Once used to it, you hardly think about it, but someone else might have trouble getting the car started.

Your start with interactive interpretation might be like one of the following two examples, depending on whether your system shares a computer with other users, or is a standalone system (just for one user at a time). The procedures described won't exactly fit your system, but they should give an idea of how it feels to get started on a system.

An "A>" (pronounced "A prompt") may come on the screen (or B>, C>, etc.), meaning that the computer is waiting for instructions. A question on the screen may ask for a password. Type your secret password, letting the computer know that you are authorized to use it. The password doesn't appear on the screen (can't have people reading it over your shoulder) but you know it registered, because a new message comes on, asking the name of your department. Look on your "cheat sheet", the scrap of paper on which you've written a few words to use on the computer. Type the expression for the name of your department, like maybe "domxpdv" for Domestic Exploration and Development, so your time on the computer can be billed to the department. Then it wants to know your name. Type that, having selected a form of it for the computer to recognize when you come back later (a short version). The computer has plenty of room for a long name, but you won't want to type your full name at the start of every session. The next question is what program you want to use. You type something like "seis2d" for "2-D seismic interpretation". O.K., that's a program it has. Now it asks what area you want. Tell it, again in the exact form it knows. Or instead of typing those abbreviated expressions, you may just use the arrow keys to select the full expression from a list, except for the password.

If you have a standalone system, the procedure might be more like this. Looking over the menu that comes on the screen, you answer one question by typing INTERPRETATION, and another with 2D. With this system that is dedicated to seismic purposes, you don't need to tell it that you want a seismic program, or the department. The menu changes to a multiple-choice type, so you switch from the keyboard to the puck. Moving it around on the desk, find where its graphics tablet is built into the desk, indicated by the cursor showing up on the screen.

You are started, the machine is running, the area is in memory, the interpretation program is called up, and you are ready to select a display.

3

Defaults, Options, Delays

The ways data can be displayed may be set in the system as defaults, standard settings that you can use if you don't want to change them.

When you are new to a system, you will probably use the defaults for most of the choices. Then, as you get more comfortable with it, you will branch out, selecting settings that better fit the interpretation you are doing, or in some cases that just are more pleasing to you. To learn what the choices are and how they affect your work, you can experiment with the various alternatives. It may be best to experiment with just one set of choices, like all the different trace spacings. Select one of them, call up a section, see what it looks like, go back to the choices, select another of them, display the section with that one, etc. Then you will be comfortable with those choices. Some other time you can try variations of another parameter. A good time to do the experimenting is when for some reason you are particularly concerned with that function. Learning about the choices then might allow you to use a better one for your interpreting purpose at the time.

The defaults in effect when you use a system may be the choices you made in your preceding session with the system. Instead of just saving your settings, your system may have a function that also saves the displays you have on the screens at the end of an interpreting session. The next time you start, it will put everything as it was when you stopped. You don't have to select and call up. You are ready to resume interpreting where you left off. You are in the same program, with the same thing on the screens: section, colors, picks, and map.

TAKE A COLOR, ANY COLOR

Some of the options have to do with color. In color schemes most people have their own preferences, either arrangements that heighten contrast for them, or colors that look pleasing to them. For interpreting reasons, you may want bright colors with high contrast to make differences stand out strongly, or pastel colors, so the differences are visible but don't force themselves on your attention. Or optical reasons may be more important influences on your choices of colors. Bright colors and high contrast may be fine for a brief session, but may tire your eyes if you look at them for a long time. A color-impairment in your vision can affect your choice of colors. Where other people would use red and blue, you may distinguish green and yellow better. If either you or someone you show your interpretations to has such a problem, you can have a color arrangement for that person, and another for other people.

Choosing colors may be a matter of calling up a standard display of many colors and selecting among them by moving a cursor to point out the colors you want (Fig. 3–1a). The colors may range horizontally from light to dark, and vertically down the spectrum from violet to red, or some similar sequence. You just position the cursor on a shade you want, and press a button.

Another way some systems use for making the choice is by moving the cursor on a simple uncolored line diagram, circle, or something similar. You place the cursor between red and orange to select red-orange, for instance, or between dark and light shades of each of three colors (Fig. 3–1b). With such diagrams, you can move along scales of hue or darkness. Each of the two methods has advantages. With the mass of color, you actually see all the colors before choosing one. With a diagram you have a better idea of where the color you select fits in with other colors, and this may allow you to get nearer to the color you want.

You can choose the colors for the commands you will be looking at on the screen: the words that will display commands for you to select

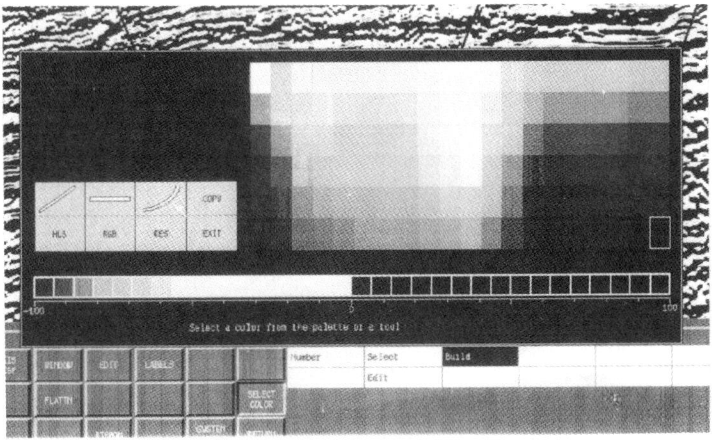

a Mass of colors *Sierra Geophysics, Inc.*

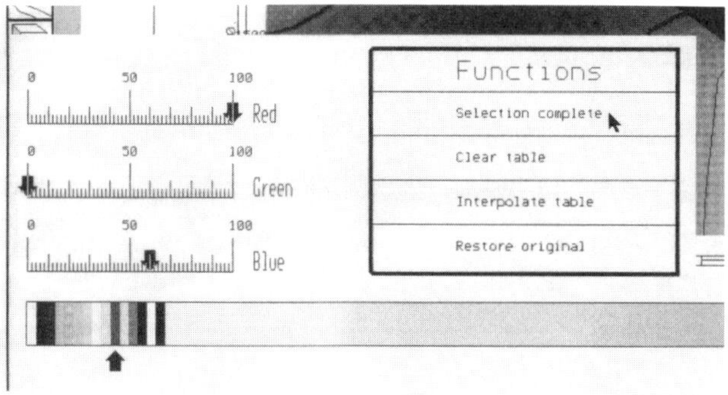

b Selection by shades *GeoQuest Systems, Inc.*

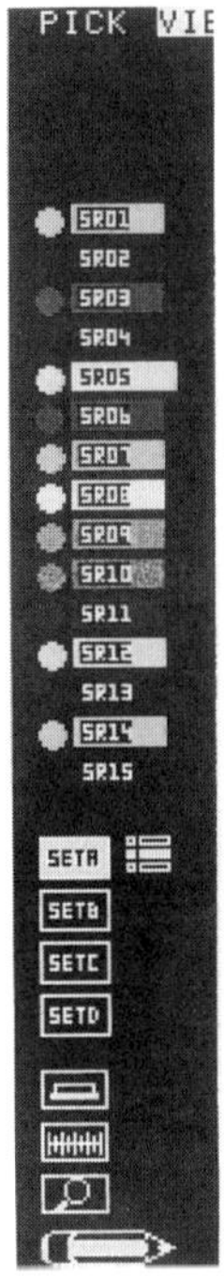

Graphic Science Corp.
c Pencil box

Fig. 3–1 Color choices

from, and background colors for the words. It's fun to play around with the colors until you get a set you like.

You can also choose colors for a section, peaks in one color and troughs in another, for instance. Different attributes of seismic data can be displayed in color superimposed on a wiggle-trace section, so both the attribute and the wiggles are visible at the same time. A time slice can be displayed to show any attribute in color.

The color can be added to a wiggle-trace section in different ways. A variable-density color section can be overlaid on the wiggle section. Or the wiggles can be filled in with colors. In either case the colors give additional information that isn't evident on the wiggles themselves.

The picks you mark on sections and slices have to stand out from the colors the display is made up of. In some systems this isn't a problem, because the picks are shown as lines of solid color that are thick enough to be easy to distinguish from the mass of gradations that make up the color section. In other systems the lines may be thin, so you have to make sure the picks are in a contrasting color. That can be difficult if you occasionally change section display colors. The color for a pick that contrasted well with the colors of one display may blend in so well with another set of colors that you can't see it. You may find yourself sometimes having to change the color of a pick, and then change it back, so you can see it on different displays.

For selecting a color for a horizon, one system has a clever way of keeping an element of the familiar in the choices with a limited number of colors displayed in separate little blocks along one side of the screen, like a box of colored pencils. You can select one of them with the cursor. Or if there isn't one you want in that box, you can request another set of colors, like opening another box (Fig. 3–1c).

As on paper sections, you will usually want to choose a different color for each horizon you are picking. After working on the interpretation a while, you will find that you have memorized which color represents which horizon, unless you changed the horizon colors from time to time to make them show up on differently colored sections. If reflections aren't easy to recognize by color, you can have the system label the horizon with its name. A problem with the labeling, though, is that there may be a label for every segment of a horizon you pick. If, because of faulting or for some other reason, you pick in many short pieces, a section will be cluttered with labels.

One of the delights of choosing colors for a section is in varying the colors and watching them change on the section. There are several ways color can be manipulated. You can insert some contrasting colors in the set of colors. You can select a range of colors, and then stretch it so the same colors are spread out over more of the parameter. Similarly, you can compress the range of colors.

For map displays, you can build up an arrangement of colors for the different reflection times. A bar near the map shows the colors you have chosen, labeled as to the values they represent, so you can refer to it to determine the approximate time or other mapped characteristic at a point on the map.

To make a color map that looks like a contoured map, you will normally assign colors to uniform ranges of times, or whatever you are mapping. But, since you can assign colors in any way you wish, you can also add a color to mark some specific contour. For instance, on a structure map, you might want to make the oil-water contact especially noticeable. To do that, you can have a narrow range, maybe only one specific time, shown in a highly contrasting color, like black. Your system may have a provision for placing such a marker. If so, it's easy to do. If it doesn't, it still isn't very difficult. Just take the time to assign the color.

In any of these uses of color, you can choose the scale of colors to suit yourself in any way you wish: to be more contrasty, more restful, or just more pleasing. Often you will want a color selection that makes whatever you are looking for stand out, but if you are working at a screen for days at a time, restfulness and esthetics may become more important to you.

TRACE DISPLAYS

One of the major functions of interactive interpretation is making the data look different so you can see different things in it. Displaying a section in different forms of traces is one way to get a number of different ways to look at the data.

The most basic choice in form of traces is between wiggle traces and variable-density ones. You most likely make the choice by moving the cursor to an option on the screen and clicking. Within those choices there are variants. The wiggle form can be shown as just wiggly lines, as those lines with peaks or troughs or both filled in, or as just the fill-in without the lines. The fill-in can be in different colors. The variable-density form can be shown in shades of grey or in any of a wide variety of color arrangements.

The display most like a conventional paper section is a black and white section in wiggle/variable-area form. These displays are occasionally used to examine in detail differences in the characters of reflections. On a wiggle-trace section, character is shown clearly in the shapes of the wiggles. The normal character of some particular reflection might be, for instance, a strong peak, a weak trough, and another strong peak. Then as the rock layer changes in thickness or lithology, the character may change to an entirely different one. For this type of study, the section may also need to be displayed on a large scale. So a system may automatically switch to wiggle-trace when you enlarge the display beyond a certain scale.

For most seismic interpretations, though, the section is shown in variable-density form. Variable density is better for interpreting on the small-scale sections that are best at showing structure. It is also good at showing stratigraphic changes.

To see a wiggle-trace display as separate wiggly lines, there must be some space between the lines. It's O.K. for the high amplitudes to overlap, but the relatively straight parts of the lines must be seen as separate. A variable-density trace, though, can be made up of just one column of dots, because it shows amplitude differences by shades of grey or of color. The next trace is shown by the next column of dots, so the two touch. This makes a very compact form of display, that can show a large number of traces on the screen.

In a paper black-and-white variable-density section, the difference between peaks and troughs is indicated by a grey scale from black to white. But the diazo reproduction process that is used to print paper sections does not have much range of greys. So the variable-density sections come out essentially in black and white, with black peaks and white troughs. In sections shown on a computer screen, a much better grey scale can be displayed. Or color can take the place of the grey scale, with, for instance, the highest amplitude of a strong peak shown in dark blue, the high amplitude of a strong trough in dark red, zero amplitude in white, and other amplitudes shown by gradations in between.

The two basic forms can be combined, the normal seismic information shown as wiggles, and some other characteristic overlaid in color as variable-area (Fig. 3–2). Or a combination can be made with wiggle-VA, in which the fill-in is varied to show some quality.

SEISMIC ATTRIBUTES

Other options let you choose among several different characteristics of seismic data that can be singled out and displayed separately. They include conventional amplitude and the "complex seismic trace attributes" which are usually just called seismic attributes. The word "complex" in the name means that they are derived from a "complex" mathematical expression, that is, an expression that contains what mathematicians call "real" and "imaginary" numbers. These attribute displays are made by eliminating some part of the other information from each of them. The information is thus split into parts that make different features of the seismic data easier to interpret. These different displays, just by making the data look different, can give you an opportunity to see things in one display that you hadn't seen in others. These varied displays are part of one of the major reasons for interactive interpretation. They make the data look different, and so allow you to see features of the data that can lead to your making a better interpretation.

These attributes may be known by different names, depending on the vendor. But they are generally known as reflection strength (or amplitude envelope), instantaneous phase, cosine of phase, instantaneous frequency, smoothed frequency, and apparent polarity. The first two are probably the most commonly used, reflection strength for bright spots and instantaneous phase for improving continuity of reflections.

Other characteristics of the data are also attributes, of course. The word just means the qualities, features, characteristics of something. But the term "seismic attributes" has come to mean the features that use certain

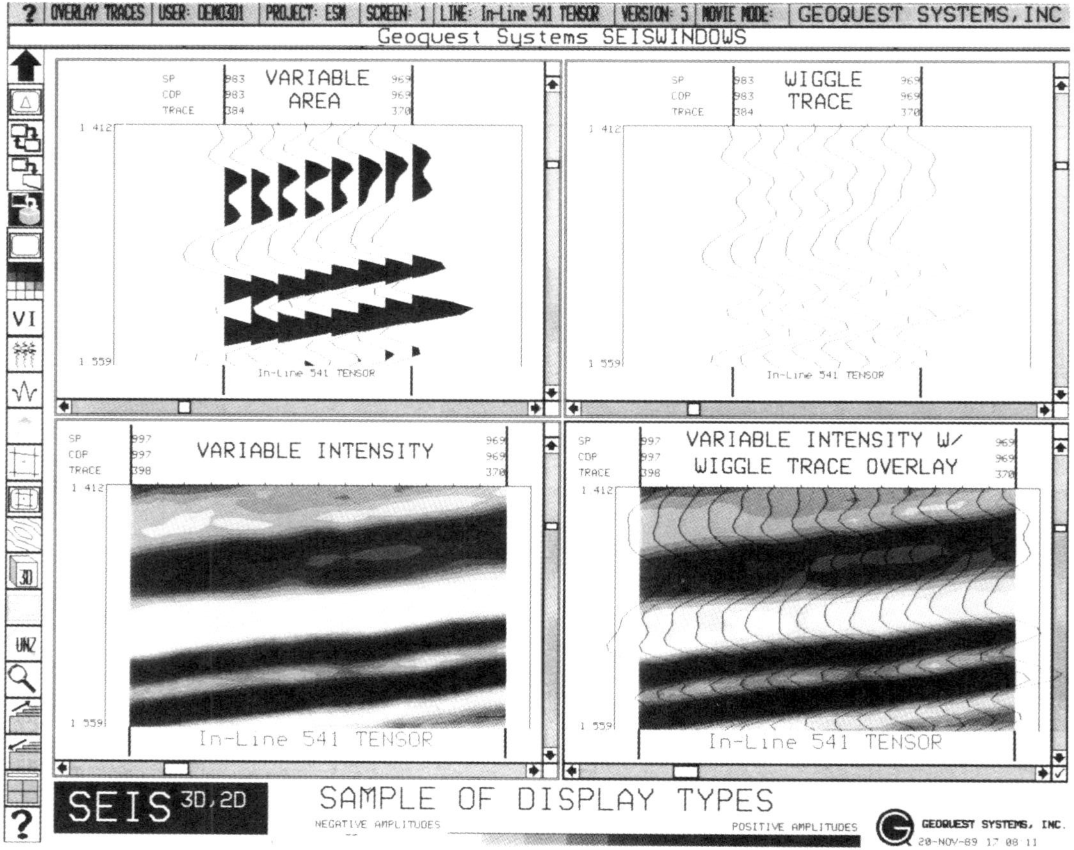

GeoQuest Systems, Inc.

Fig. 3–2 Forms of trace display

mathematical manipulation of the data to enhance some characteristics while eliminating others (Fig. 3–3).

Ordinary amplitude is not considered one of the seismic attributes. It can, of course, be seen in a wiggle-trace section as the magnitudes of the peaks and troughs. And in a black-and-white variable-density display, it is shown by the shades of grey. But it can be made more evident to the eye by adding color that varies with the amplitude. Different color schemes allow different aspects of amplitude to be stressed (Fig. 3–4). These differences in color are another way to make the data look different, just as the attributes do. Variation of color patterns can also be used on the attribute displays. That gives you a tremendous range of appearances for the data, to enhance different features.

Probably the first practical use of color reproduction in seismic sections was for some of the seismic attributes. The color processes of the time were slow and expensive, and the color sections had to be ordered from the seismic processor. So a color display wasn't made lightly, or at a moment of inspiration. Now, though, a color attribute section can be made in a few minutes, or, if it is already made and stored on disk, it can be called up as quickly as any other section. And a color print of it can be made in a few minutes, just by giving a command. Color prints may still be somewhat expensive, depending on the color plotter used. It is often better to just call the sections up and look at them on the screen. But there are a number of reasons for making color prints. If your time on the system is limited because of taking turns with other interpreters, or because you are renting the system by the hour, then it may be worthwhile to make some color hardcopy of attribute sections to study at your leisure back at your desk.

TIMING LINES AND TIMES

A vital part of a paper section is the set of timing lines running across it horizontally, like the horizontal lines on sheet music. The lines are necessary to keep your place in picking. They are labeled with round-number times, so you can quickly see approximately where you are on the section. And after a section is picked, you use them to read the times of the picks, for posting on a map.

With an interactive system, timing lines aren't so necessary. You can easily get along without them. The system will display them overlaid on the sections if you want, but it may be more convenient to not have those rigidly horizontal lines leading your eye into making flat picks. There is a function that tells you where the cursor is. This location indicator shows, somewhere on the screen, a small group of numbers, giving line number, trace number, reflection time, and amplitude. So at any time, you can tell

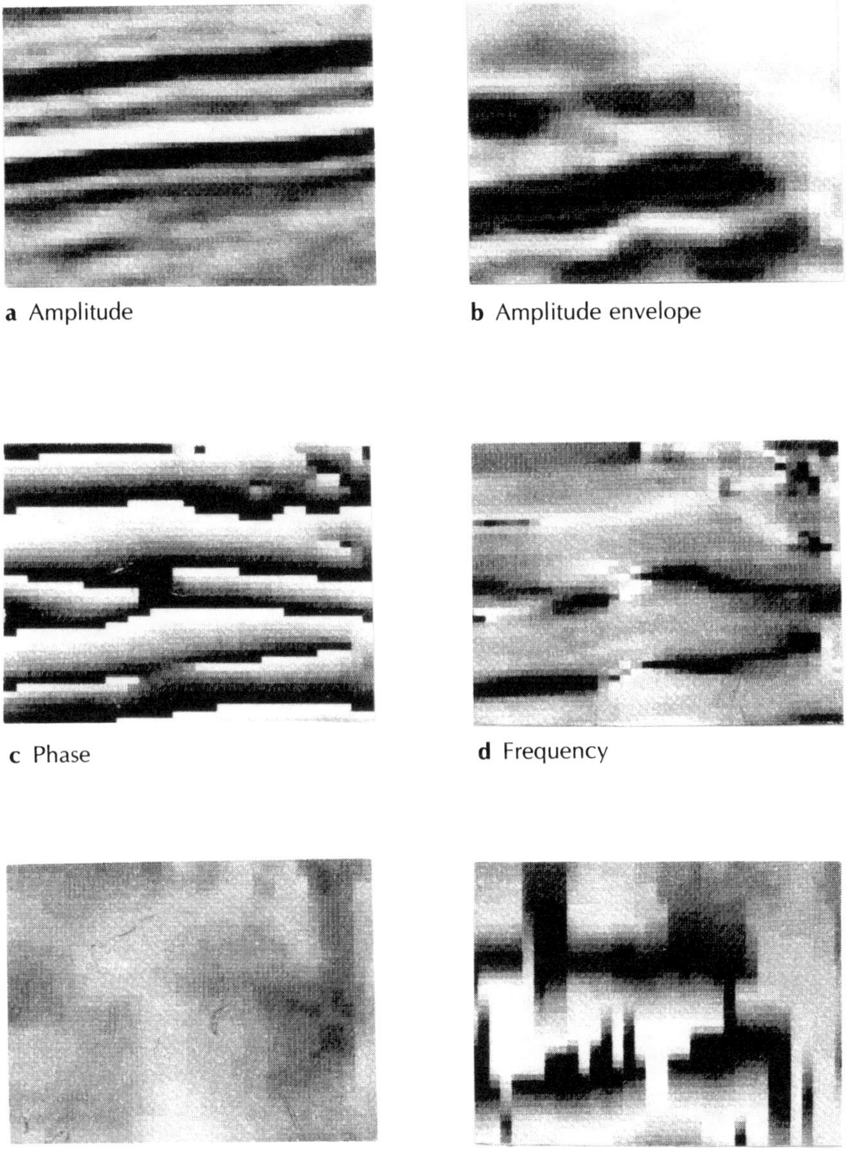

a Amplitude

b Amplitude envelope

c Phase

d Frequency

e Smoothed frequency

f Apparent polarity

Western Geophysical Division, Western Atlas International, Inc.

Fig. 3–3 Seismic attributes

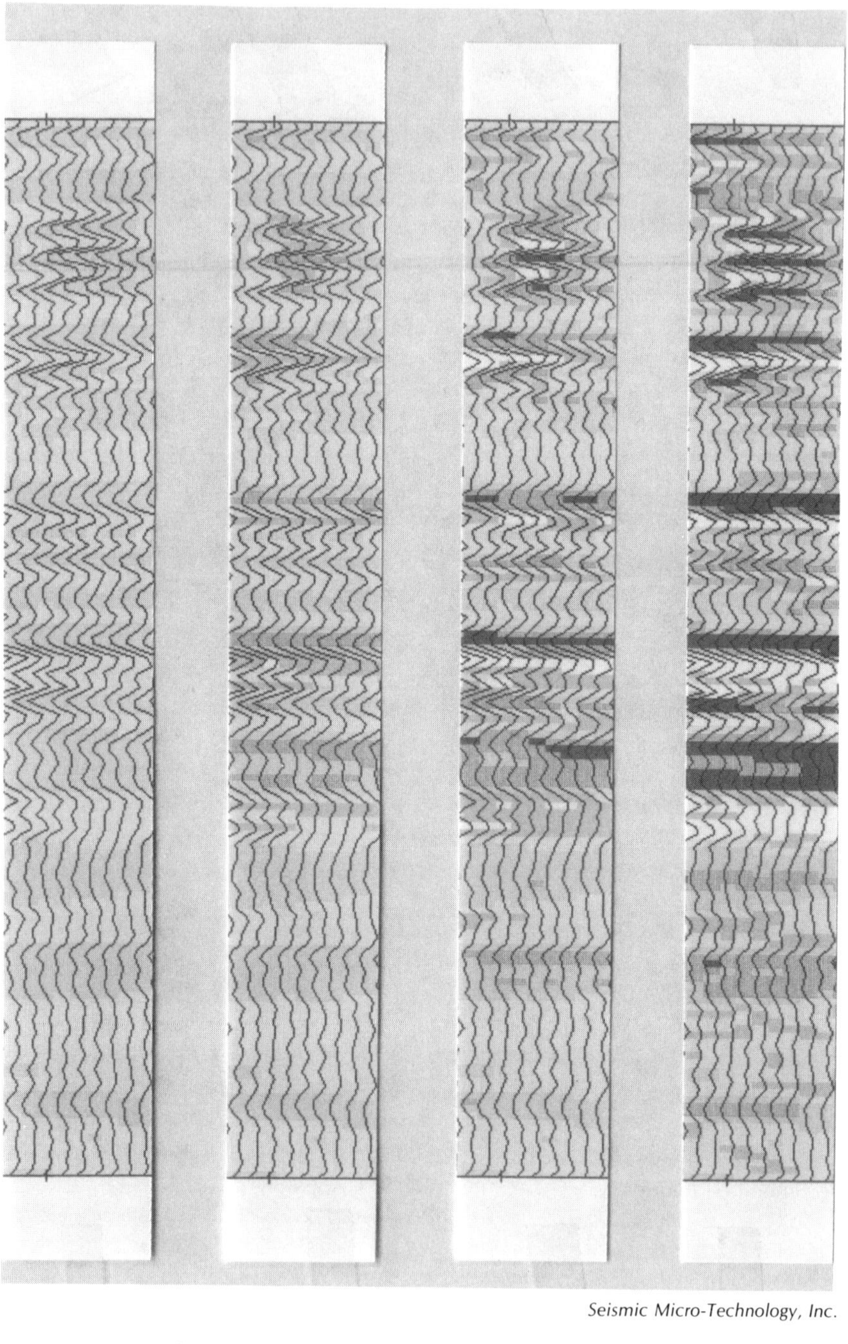

Seismic Micro-Technology, Inc.

Fig. 3–4 Variation by color (see in color on cover)

exactly where the cursor is. The information may stay on the screen, with numbers changing frantically as you move the cursor here and there on the section or slice. That type of display is usually in some out-of-the-way place, like along the bottom of the screen. You can glance down whenever you wonder where you are. Or the display may be shown only when you request it by pressing a button (you couldn't call it up by pointing with the cursor, or you'd have moved the cursor away from the spot you wanted to know about). Then it immediately shows up, not out of the way at all, but prominently somewhere near the cursor position, and maybe with an arrow pointing to that position.

The location indicator keeps you from getting lost on the section. And you don't have to read times of your picks at all; the computer reads them automatically. So you don't need timing lines, and may prefer not to have them. But if you do like them, to show you which way is horizontal, or just for old times' sake, they are available. You can have many lines or just a few, make them thick or thin, and choose which ones to label with their times. And, of course, you may need them on hardcopy, which can't have such a location indicator for every point in the print.

Time labels along the edge of the section on the screen are another matter. They are convenient for deciding what range of times is displayed. They let you see about where a feature is on the section, without having to move the cursor to that position. They don't get in the way of picking. So they are almost always displayed at the edge of the section. Similarly, shot point numbers and/or CDP numbers are displayed at the top of a section. On a time slice, line numbers are displayed, inlines along one edge and xlines along another.

Another aspect of timing is in requesting a section to be displayed. Since the screen isn't as big as the usual paper section, then to see as much detail as you'd see on paper, you need to cut down the amount of the section to be shown at one time. In picking, you are normally interested in less than the whole vertical extent of the section. So you need to show just some mid-range that includes the reflections you will want to pick, and maybe some more features. You might find it useful to look at, say, the part of the section from about 1.8 seconds to around 2.4. A system will have some way for you to tell it that, maybe a column of numbers representing times, so you could run the cursor to 1.8 and click, then to 2.4 and click again. The column may be invisible except for the number the cursor happens to be on, or the whole column may remain visible. Or a system may have a vertical bar at the side of the screen representing the whole vertical extent of the section, and you could mark a part of that with the cursor. Or you could enter the times by putting the cursor on the digits of a keypad image, or even punch keys on a keyboard.

BIG AND LITTLE, SQUEEZED AND STRETCHED

With interactive interpretation, there is great freedom in scales and proportions. Sections, slices, and maps can be about any scale you wish. You can have a small-scale section to look at the larger features, rather like hanging a paper section on the wall and looking at it from across the room. Or you can display a section in a large scale to inspect some detail, like peering closely at a paper section or using a magnifying glass on it.

You don't have to plan ahead to have the scales and proportions you want. By contrast, working with paper sections, you have to be concerned about the sizes and proportions of sections before they are played out by the processors. Although one disadvantage of interpreting interactively is that the screen isn't large enough to hold a large, long section; even with a paper section you have to either look at a limited amount at a time or back off for an overall view.

On a screen, the overall size of a section is already pretty well settled, to nearly fill the screen with section. The scale is determined by you. First you choose horizontal and vertical scales, then select the starting and ending times and the starting and ending shot points you want displayed. As long as displaying that amount is within the system's capabilities and fits your specifications of scales, the section you indicated is displayed. If it won't all go on the screen, then part of it is shown, and you can move the display to other parts when you need to.

But you aren't stuck with that scale. Any time you want a different scale, you can get it. Usually you would be enlarging it to see more detail (Fig. 3–5). There are various ways that this is done.

A hardware zoom, also called pixel replication, enlarges without adding detail. It has the advantage of being quick, so you can call up a zoom at any time without having to wait for a slow forming of the enlarged section on the screen. The image on the screen is composed of pixels (PICture ELements), colored dots that are normally so small you don't see them individually. There is one pixel for each sample, so each one represents 4 ms. or whatever the sample interval is. Pixel replication just adds more dots of the same color to each original dot. If there is one blue dot, and you ask the computer to double the scale, it puts three more blue dots with the one, to form a blue rectangle twice as wide and tall as one pixel. Adding more pixels enlarges further. But the effect is like enlarging the pieces the picture is composed of. After several stages of enlargement, the image looks crude, made of large rectangles of color. The picture takes on a blocky appearance. You would normally use this method to get a quick look at a little more detail. But you usually wouldn't want to enlarge many times, or your picture would look like it was made out of bricks.

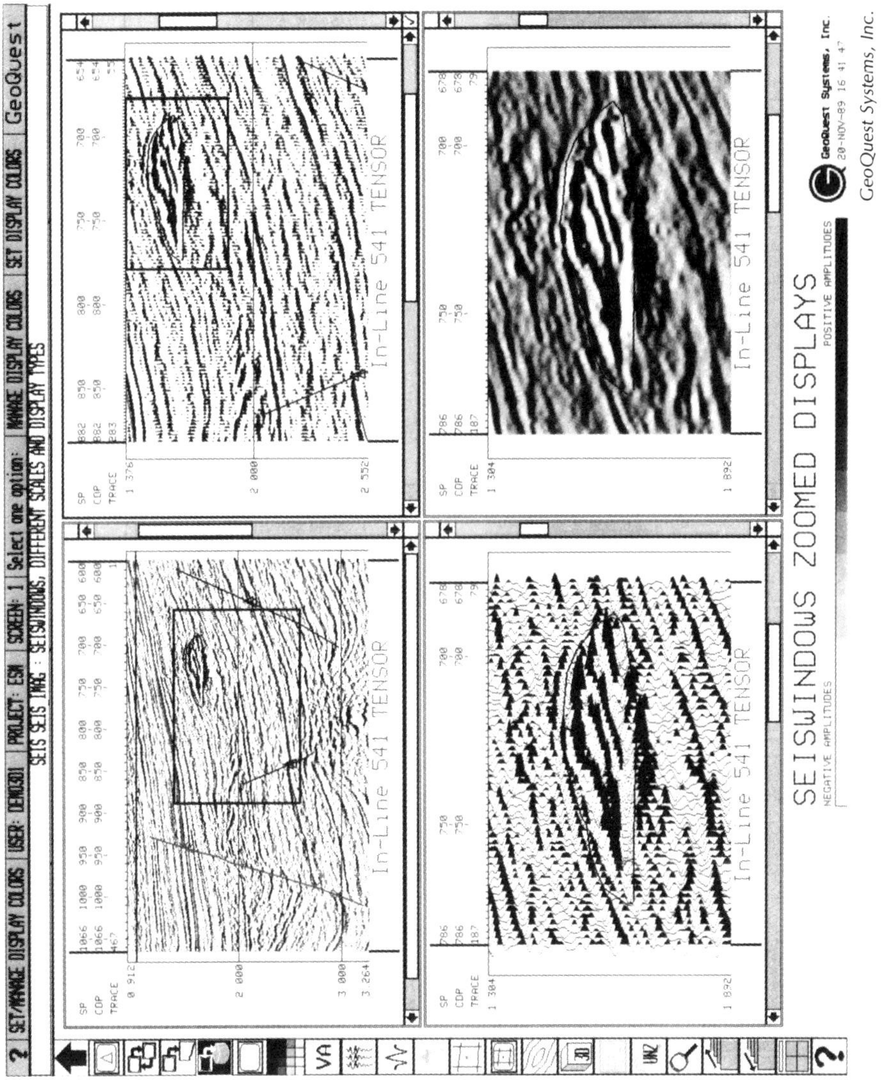

Fig. 3–5 Zooming in stages

The other kind of zoom, a software zoom, is accomplished by the program. It can calculate more detail and display it by making adjacent pixels different, rather than copying pixels. That is, the added pixels can be interpolated to be similar to the pixels around them. The effect is of seeing more detail when the picture is enlarged. A disadvantage is that it forms more slowly than a hardware zoom. Another is that in calculating the extra detail, the computer may put in detail that isn't real. If the data was sampled every four milliseconds, then enlarging to show detail every millisecond can be done only by estimating. However, some systems do such good interpolation that the effect is as though the data was recorded at a closer sample interval.

People speak of enlarging a part of a display to show more detail as zooming up, and of reducing the scale to show a larger part of the section as zooming down. Or they may, like television announcers, call the enlarging action zooming in (as in "We'll zoom in for a closer look.") and call the reducing zooming out.

Some systems zoom only in steps of doubling or halving the scale of a display. To zoom up, enlarging a part of the section, put the cursor on the command, often a magnifying-glass icon, and press a button. A rectangular outline with the cursor in its center appears on the screen. By moving the cursor, drag the outline to where you want it, and click. The image on the screen disappears and is replaced by one at twice the scale, with the spot you chose at the center of the image. Zooming down just undoes the zoom up, not giving you a new choice of centers.

In this zooming up and down in two-to-one steps, if you want an image at four times the scale, you have to zoom up twice. That means waiting for the double-scale image to be completed just so you can tell the system to double it again. That usually isn't too inconvenient, as you often may not know how much enlargement will best show the feature you want to see. Also, in zooming up, each successive image requires less time because the computer isn't having to handle as much data. But after zooming up, you will often want to zoom back down all the way. Again this takes the intermediate step, and this time each successive step takes longer. A system may have a command that stops a display, so you can go on to the next step without waiting for the display to be completely drawn.

What if you would like a display enlarged to one-and-a-half times the scale on the screen? A doubling or halving wouldn't do it. There is an advantage in that, though. If you could enlarge in small increments, then you might find it hard to get back to a scale you had used earlier. With just doubling and halving, it's easy to decide what scale you used earlier, as there aren't many possibilities to choose from.

Other systems, though, may zoom in fine increments. You give the command to enlarge, click on two points on the image, a rectangular outline forms on the screen with diagonally-opposite corners on the two points, and the system fills the display area with the part of the section in the rectangle.

The relative horizontal and vertical scales are another matter of concern. Some people, when working on paper, like a section that is stretched horizontally, so reflections look fairly flat. That lets them pick with a straightedge, and leaves the subtle features to be seen later, after the picks have been posted on a map and contoured. Geologists often like to have a proportion (of horizontal distance to vertical time) that gives the section about the proportions (of horizontal distance to vertical depth) of the subsurface.

Most geophysicists prefer to work on a section that has some vertical exaggeration. For some purposes it is useful to have even a highly squeezed section, to make subtle features stand out prominently. But then, if you are working on paper, it is hard to be sure you are directly below the marked location of a shot point, so you can pick a horizon at that shot point. Interpreters using paper sections usually settle on one or two scales and one proportion, because having all the sections in several forms would be costly and would make the storage and paper-handling problems worse. With an interactive system, you can look at whatever proportions you want (Fig. 3–6).

Compressing a section horizontally is a useful tool for several purposes. It provides vertical exaggeration that makes subtle features easier to see. It can make it easier to detect the flatness of flat spots. The compression doesn't enhance the flat reflection itself, but by increasing the steepness of other reflections around the flat one, it makes the difference stand out. Faults generally differ from reflections in being steeper. Compressing makes the faults look nearly vertical, so they too stand out (which may make it easier to pick them). Also the breaks in reflections at a fault are mashed together so they are more easily seen to line up. Compression also lets you look at more of a seismic line at one time.

There are two ways that interactive systems compress sections, by plotting traces closer together and by decimation of some of the traces. The simpler method is having the traces close together, but there is a limit to the degree of compression you can get that way. A trace has to be at least one pixel wide to appear at all, and there are a limited number of pixels in the width of the screen. Also the display uses some space for annotation, so the pixels available for a section are less than will fill the screen's width. To compress beyond that limit, systems use decimation, eliminating some of the traces. That is quick and expedient, but it has a disadvantage. The skipped traces may have some of the information you need to see, so leaving them out may cause aliasing. If there are frequent

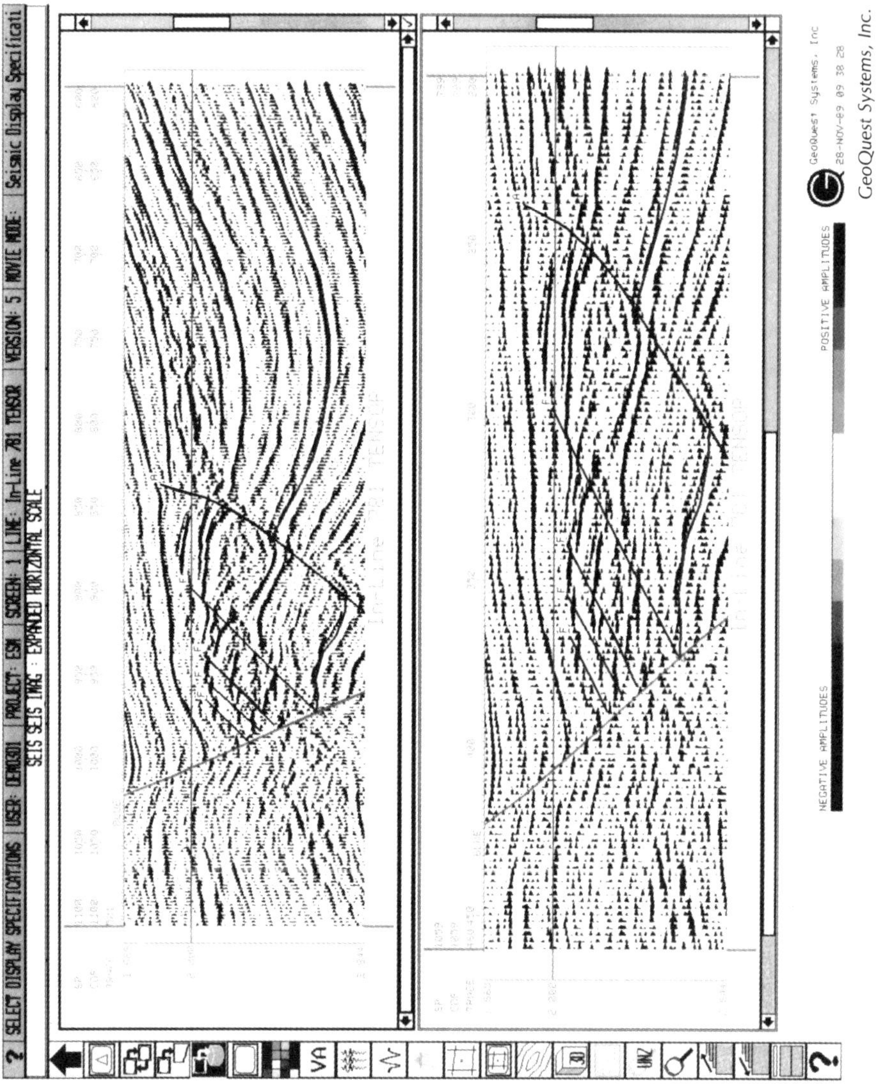

Fig. 3–6 Horizontal stretching

small features, and many traces are dropped, then a feature could be skipped, and not seen at all. Or the area between two features could be omitted, making the two look like one larger one.

In one way around that problem, although the decimation takes place on the screen, if you are having the system do some degree of automatic picking (we'll get to that later), the system may pick all the traces, including the ones you can't see. So when you later look at the section without compression, the picks include those features you hadn't seen, but only if the automatic function was able to pick them correctly. Anyway, the problem usually isn't severe, because you normally use the decimated displays just to see the regional picture.

So in both scale and proportion, you no longer have to decide on a compromise, and live with it all through your interpretation. You can change either of those parameters whenever you want to, for a quick look or to continue using that type of section. And when you make a pick on one type of display, it will be shown in its correct location on any other type. There is no need to copy a pick from a compressed section to a normal one, for instance. It's already in a form to be displayed on either.

But these handy screen displays don't have as much detail as we might like.

SCREENS AND DETAIL

One of the worst problems with interactive interpretation is screen size, or rather, a combination of size and detail. Seismic interpreters are used to spreading out long sections and looking along them. But when you start interpreting by computer, you have to settle for a smaller area of section, with less fine detail than you are used to seeing.

The problem is partly compensated for by ease of calling up a different part of the section, and by scale change and compression. Instead of moving your head close to the section or getting a magnifying glass, you can display part of the section enlarged. Instead of looking back and forth on a long section, you can have the different parts come onto the screen. And instead of squinting to foreshorten the section and make low-relief features more evident, you can display it compressed, so the features look more pronounced.

But the amount of compression you can use is limited. If you are displaying the section in variable density, you may already have as many traces on the screen as it will hold. Each trace requires at least one column of pixels, and the screen has only a limited number of pixels. If the display is in a wiggle-trace form, each trace takes up more room, so the section can't be compressed as much. Sometimes people even squint at a display on a screen for additional foreshortening.

A home television screen has about 500 dots each way to make up its picture. The pictures you see on it are mostly in motion, so your eye doesn't have time to dwell on any lack of detail. The screens used for interactive interpretation have more dots, about a thousand each way, so the detail is finer. But the pictures you look at on them don't usually move like television pictures do. In interpreting you mostly keep one section on the screen at a time, and examine it closely.

It would be easy to get a bigger screen. People often use large screens or projection screens to get large images for training or demonstrations, situations in which a number of people need to see the screen at the same time. But a larger screen doesn't add detail. Like a larger television set, all you get is a bigger image, an enlargement of the one you had, with no more detail. Instead of more pixels there are bigger pixels. It's like enlarging a picture from a newspaper. The picture is in half-tone, made up of separate dots. Enlarged, it can be seen from farther away, but the same dots are there, just bigger.

One reason that a screen doesn't use more dots to make up its images is that every one of those pixels has to be transmitted to the screen by the computer. Twice as many in each direction would be four times as many altogether, so the computer would have to generate four times the information, the cable to the monitor would have to handle four times as much, and the monitor itself would have to display four times the information. And the goal isn't to just double the number of pixels in each direction, but to increase it considerably more than that, enough to show large sections in the detail provided by paper sections. That will probably take a different approach, not just more power, but some clever idea that gets around the problem.

But the convenience of calling up different displays far outweighs this one limitation. And remember that anything you use for any purpose has limitations. Good old paper sections have plenty of limitations. Much of learning to interpret on them is learning ways to keep track of them, mark them without making it too difficult to see the reflections, etc.

MORE SECTIONS THAN ONE

Most interactive interpretation is done with one section or time slice displayed on a screen at a time. You will often wish there was room to display more of it. But at other times your main need will be to compare sections. That calls for having two or more displayed at once. They will necessarily be even smaller pieces, but you will be able to look from one to another.

You might want to compare two parallel lines, in picking a reflection or a fault. You might want a number of sections in sequence for the same purpose. You might want two or more displays of the same section, in different display forms.

In some systems, the way of displaying more than one section at a time is related to the way a line made up of segments of lines is displayed. A section can be made of parts of several 2-D or 3-D lines, or made from traces in any direction through a 3-D area. To display several parallel lines, you would just use the same commands, except that the pieces of line would not be connected. They might be displayed as though they were connected, so you would have to be alert and realize that the apparent line intersections are not intersections at all, but just places where one line ends and another, unconnected, line begins.

A system may have a way of displaying those parallel segments separately, either one above another or side by side, but with gaps between them. Instead of several parallel lines, one line may be shown in several forms (Figs. 3–7 & 3–8).

ROWS OF SECTIONS

One way of showing more than one section on the screen is to break the screen into rectangles. A system may have a provision for showing

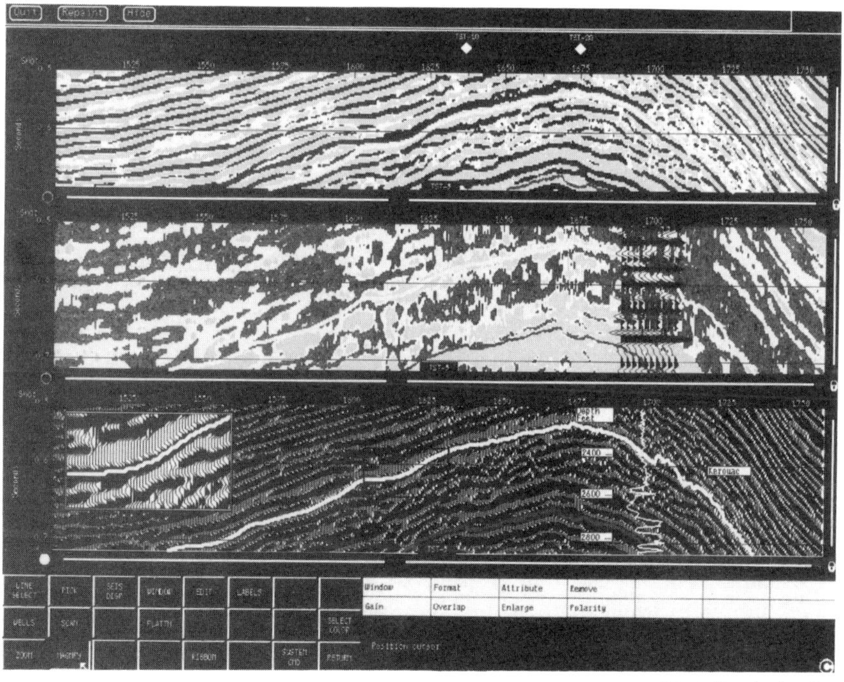

Sierra Geophysics, Inc.

Fig. 3–7 Sections one above the other (see in color on cover)

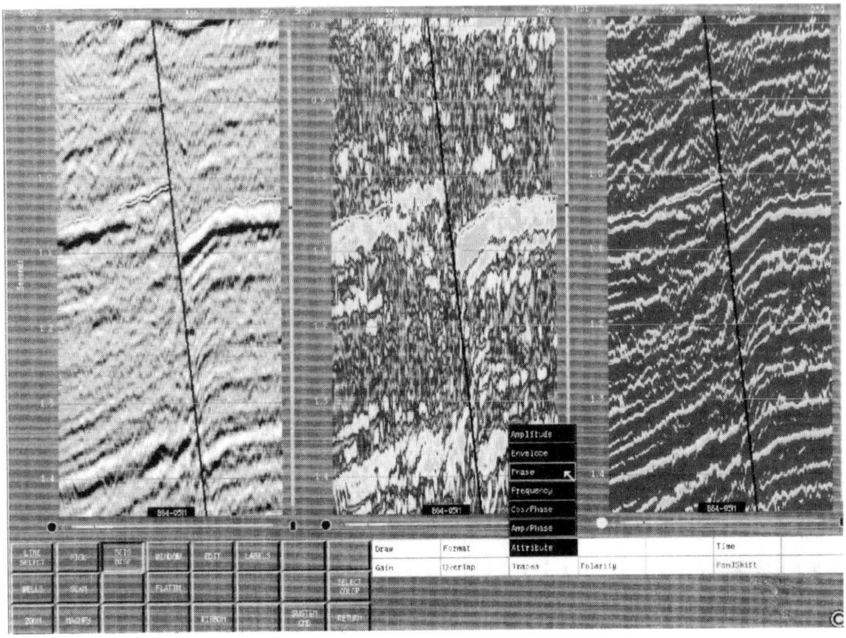

Sierra Geophysics, Inc.

Fig. 3–8 Sections side by side

one section, two, four, nine, or twelve (Fig. 3–9). Two are shown one above the other, and the four and nine are in columns and rows. It's easy to switch from one number of sections to another, probably by clicking on an icon. In 3-D, the process of filling the rectangles with sections may be automated, so you call up the display you want in the first rectangle and then give a command for the computer to fill the rest. It does so, showing the same part of each section. You can pick a reflection on one, then another, go back and change the first, and so on. Then you can call up a different one in the first rectangle, and again tell the system to fill in the others. You might start with the last one of the preceding set, or, to help you solve a subtle problem, you might prefer to start with one within the first group, so several of those you had already picked are displayed again.

Some systems use that display of a screenful of sections (or time slices) shown in small scale as the basis for quick manipulation of sections. A group of sections are called up from disk and stored as one display, so calling up any part of it is something like displaying part of a section that you already have called up. The sections in the group are shown in columns and rows on the screen. The number of sections in the group is a number

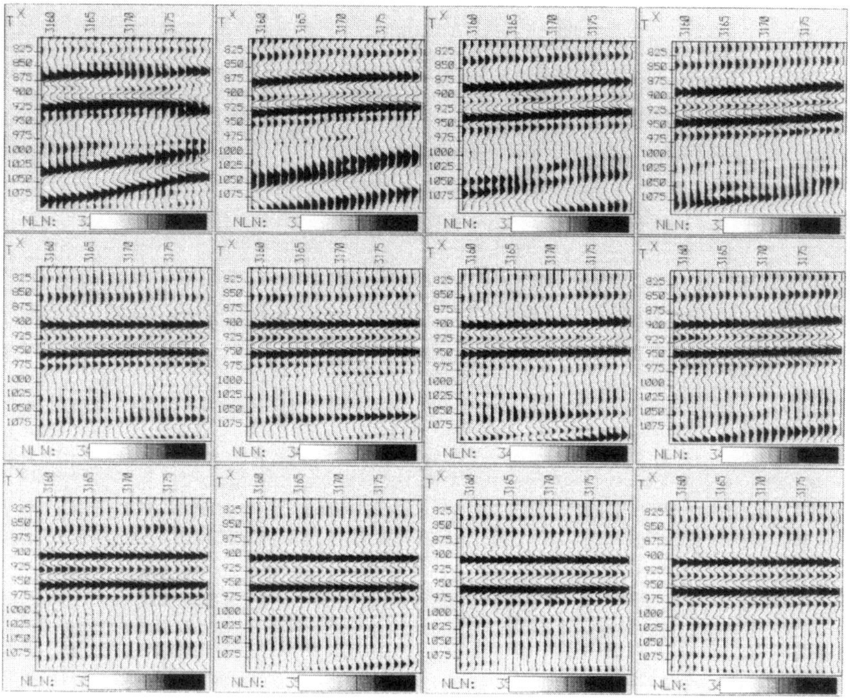

a Twelve Sections

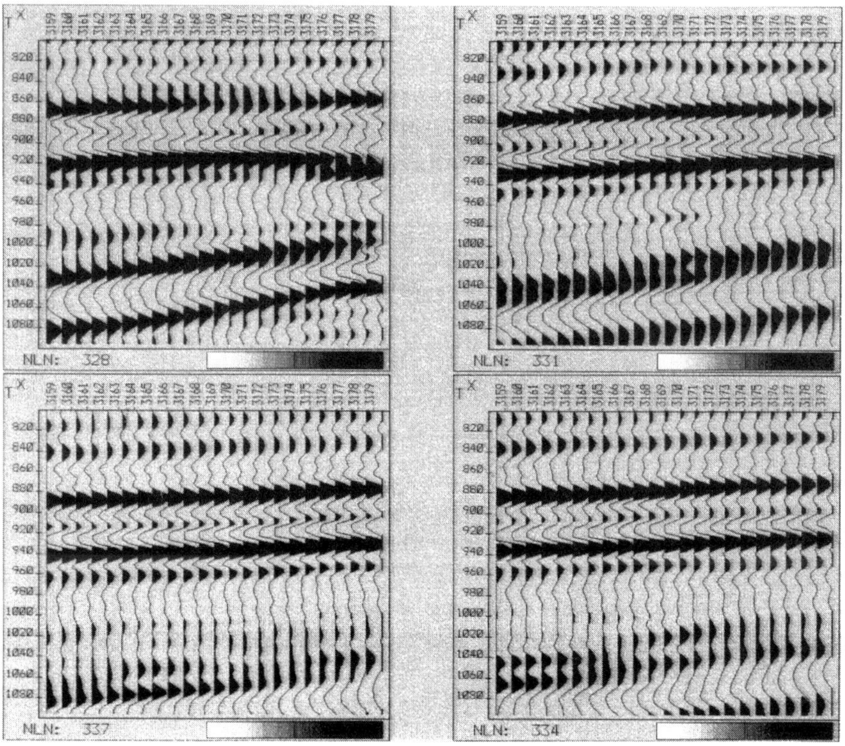

b Four Sections

Western Geophysical Division, Western Atlas International, Inc.

Fig. 3–9 Columns and rows of sections

that can be displayed that way, maybe twelve or sixteen. Any one of them, or a smaller number, like four or nine, that also fits easily in columns and rows, can be displayed quickly. Displaying fewer than the whole group, for instance, just four out of a group of twelve, or just one, is for the computer just a matter of enlarging that much of the total. Handling the sections in groups avoids the need for the program to search through the entire mass of data each time you want a section. That makes it easy to display a "movie" of just those few sections. When you've interpreted all the sections in the group, you will need to create another group.

CLUTTERED DESKS AND WINDOWS

One function that makes computer displays more like the papers people work with is windowing, the use of "windows". In writing or any paperwork, you tend to accumulate collections of papers on the desk or other surface you are working on. You look up a reference and bring it to the desk to use it. Then you track down a note you made, and put it on your desk. The most recently used papers get on top of other papers. Later, if you want one of the early ones, you look to see if any of it shows, peeking out from under the others. If it does, you pull it out. If not, you start moving papers. That works pretty well, except for the time wasted in hunting.

In its usual use, a computer screen is like a desk that can have only one paper on it at a time. But when the system has windows, one display can cover part of another. Having several windows gives the computer screen an accumulation like you get on a desk. The appearance is like a desk with papers scattered over it, some partly hidden by others. To look at one of the partly hidden displays, put the cursor on a part of it that is visible, and click. Presto, that display is shown completely, as though it was the top piece of paper.

Some interactive seismic interpretation systems use windows the same way. The advantage of windows in seismic interpretation is to let you easily compare different displays. You may have a section on the screen, and want to examine part of it more closely, by showing a window with that part enlarged. Having it on the screen at the same time as the normal-scale section lets you compare the two readily. Windows can also show the section in other forms and with different attributes, or they can contain parts of other sections, maps, etc. (Fig. 3–10).

ANNOTATION

Your system probably has an automatic annotation function. If you make it active, then each horizon is labeled on your sections (Fig. 3–11), maybe on each segment between faults and other breaks. When you pick faults and name them, they too can be labeled with their names. Similarly,

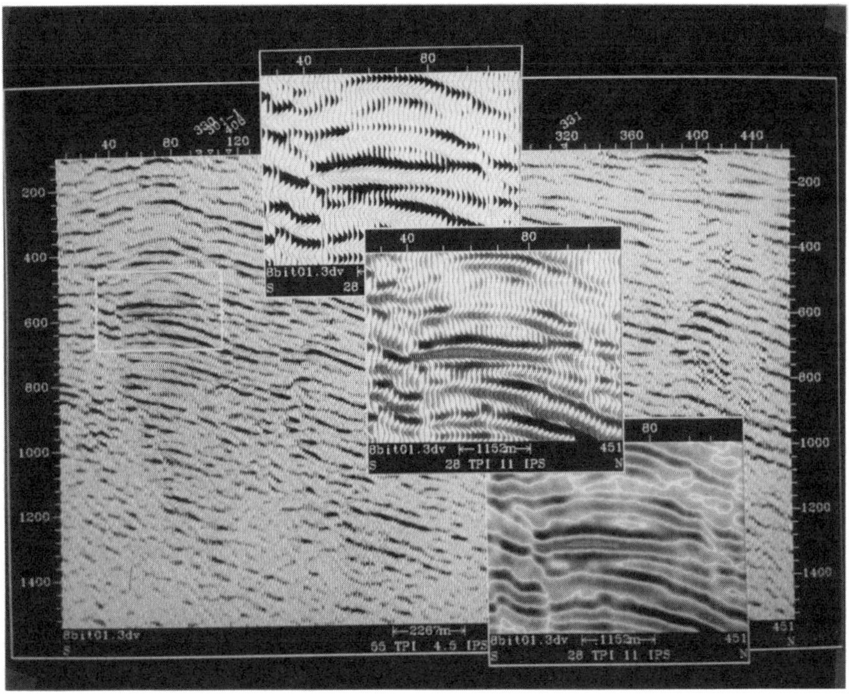

a Section

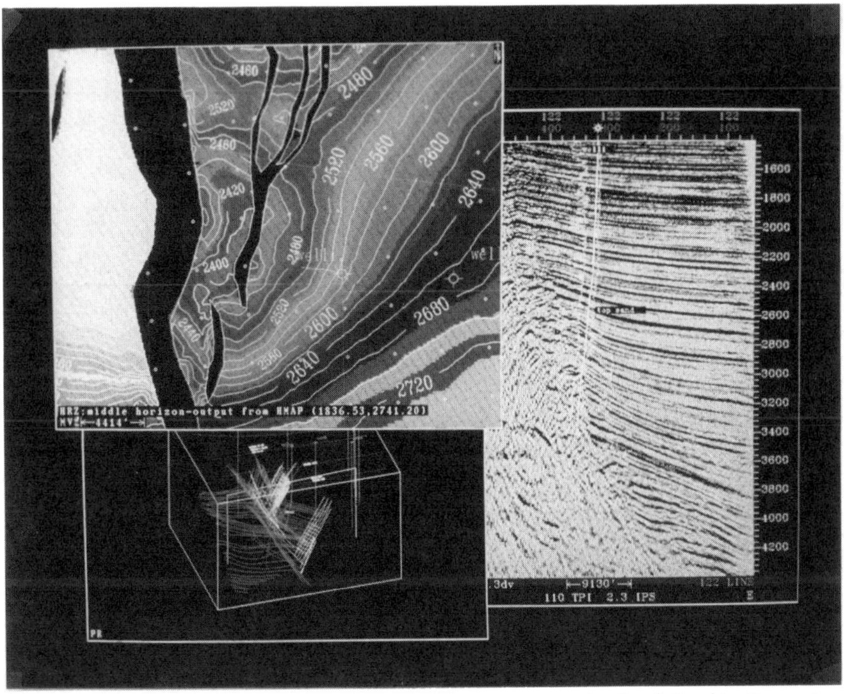

b Several kinds of display

Landmark Graphics Corporation

Fig. 3–10 Windows

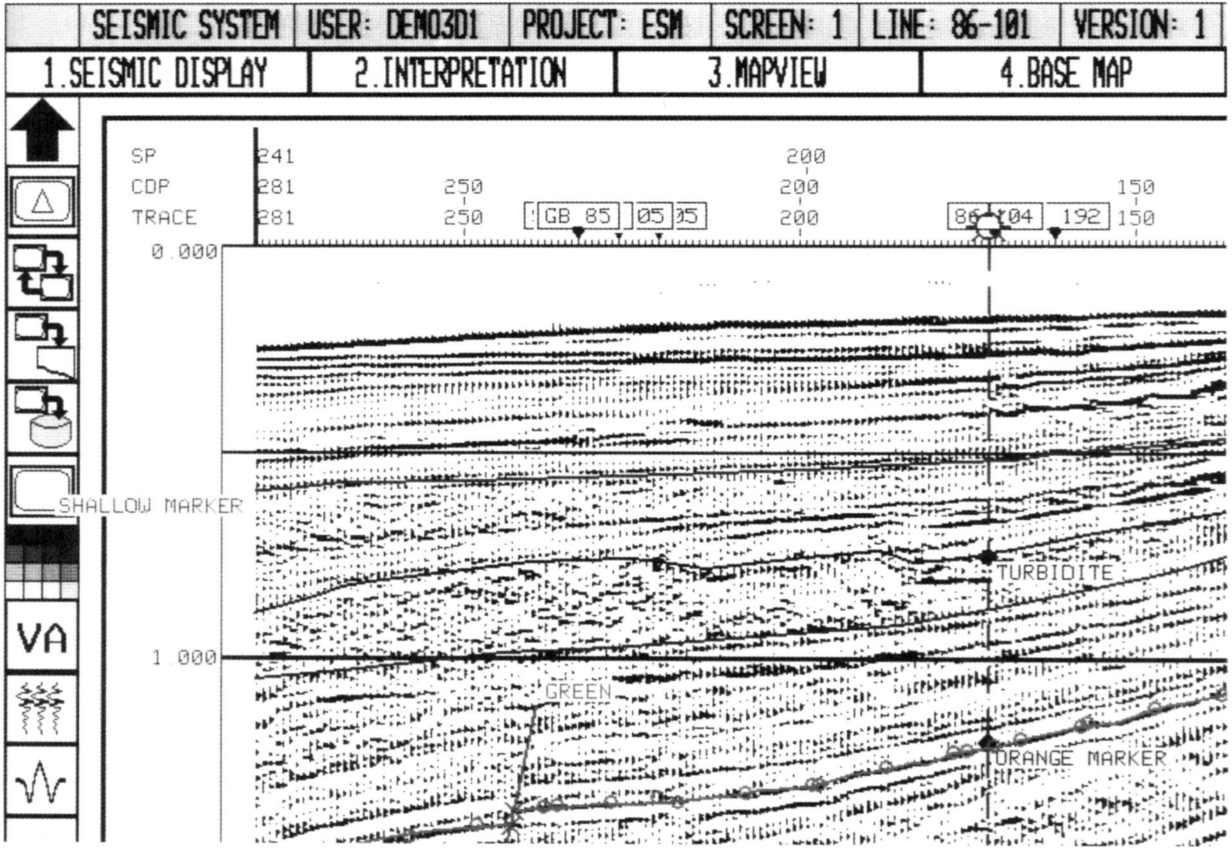

GeoQuest Systems, Inc.

Fig. 3–11 Annotation on section

where well logs are inserted in the sections, the formations can be labeled in the wells. In areas with steep dips and large faults, the labeling can help you keep track of your interpretation. In less-contorted areas, you may prefer to not display the labels, especially those for horizons. In that case, the colors of horizon picks may be enough to let you keep track, and the labels may be in your way. Then you might display labels just for formations in wells and for individual faults.

That takes care of labeling things the system has in memory. But there is another kind of annotation you may sometimes want to do—pointing out interesting features. Working on a paper section, you might write a note on the section, or better, on a copy of it. The note might be in bold lettering, perhaps with a big arrow, maybe with a bold marking of a reflection. The wording would be something like "Reef?", "Watch out for this on other lines", or "Example of multiple interference". The copy might be stuck on the wall, given to the boss, or kept in a reminder file.

In interactive work, you could just make a hardcopy and then mark on it, but that isn't very satisfactory. The paper that hardcopy is made on is shiny and might not take marks very well. And anything you drew on the hardcopy wouldn't be stored in the computer for future reference, so you'd have lost one of the advantages of letting the computer do your filing.

Your system may have a special annotation command for such notes, letting you type on the display and maybe add arrows. You just give the command, put the cursor where you want the lettering to begin, and type on the keyboard. You probably won't want the annotation on the section you are interpreting on, so you will probably use the "Picture", "Slide", or whatever it may be called, function that sets aside a copy of the display in a special file. Then you can later call it up at any time to show to someone, or to make a printout for a report.

WHICH POLARITY CONVENTION?

There are two polarities in which seismic data can be displayed (Fig. 3–12). The polarities are conventions for which direction of change in velocity is to be shown as a peak and which as a trough. When sound travels from one rock layer to the next below it, some of the sound is reflected back to the surface—but only if there is a difference in the velocity of sound in the two layers. The difference can be from slow to fast or from fast to slow. In one polarity, the peaks represent changes to a faster velocity, that is, sound travels faster in the lower layer than in the upper one. In the reverse of that polarity, peaks show where the change was to a slower velocity. A peak is usually filled in with black on a wiggle/VA section. In general, compaction of rock makes velocities increase with depth.

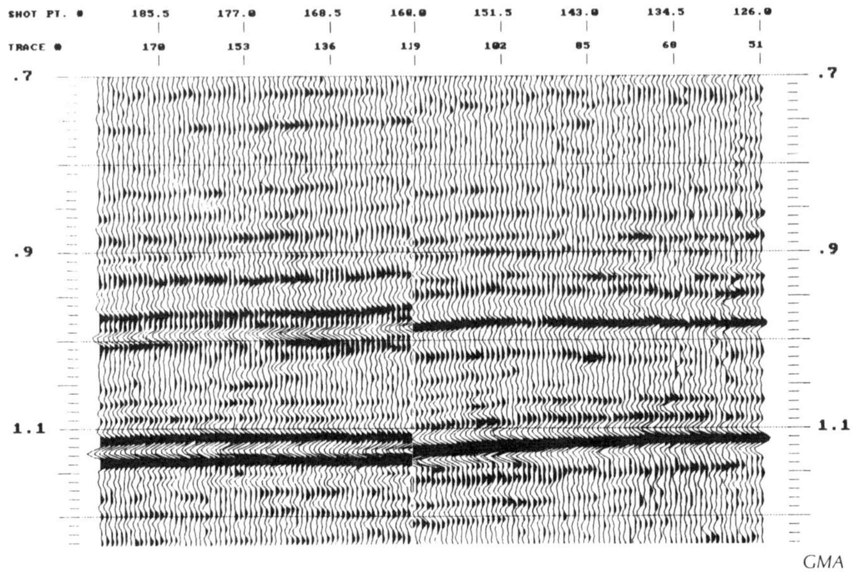

Fig. 3–12 Comparison of polarities

Peaks are usually thought of as more prominent than troughs. Peaks are conventionally filled in with color, and troughs may or may not be. Intuitively, it seems that the more outstanding mark should occur where a deeper and harder material is encountered. Thus a sea bed or the top of a limestone layer under softer rock would be marked by the black band of filled-in peaks. But people in research and processing often prefer the opposite convention. So there isn't a standard for the industry. One company will have one convention, and another company will have the opposite one.

That means that you have to be alert to the polarity of a section. Even though your company may have adopted a standard, you will occasionally work with an area that is the other way around. That may happen with trade data, or with your data when the processor misunderstood, or when your company is a partner in a joint operation and the operator has a different convention from yours.

How do you find out what convention was used? You can look up documentation. If you have paper sections from the area, there will probably be a notation saying what polarity convention was used. But be careful—the wording may be confusing. There may be an indication in the header material on the magnetic tape of the sections; you could ask the person who loaded the data about that. There might even be a notation on the screen at some stage in your calling up the area to work on.

If you can't find out which convention was used from documentation, there is still the data itself. You can check a strong, well-known reflection. If there is a hard layer that is easily recognized over the area, then the strong event at its top is one that is at a change to a faster velocity. Or a flat spot from a known gas zone, since it is at the change from gas-filled rock to liquid-filled, is also a change to a faster velocity. A reflection at the top of the gas trap is at a change to a slower velocity. One or more of these clues should enable you to determine which polarity convention was used.

Working with offshore data, call up a section and look at the first reflection on the section, the one from the sea bed. If it is clear-cut, then whichever of peak and trough is the first strong leg at the sea floor is the one that is used for a change to a faster layer. The sea bed may not be easy to detect though, either because the bottom is mud-covered and doesn't yield a strong reflection, or because the shallowest part of the section has been slighted in processing.

Handling data on your company's workstation, once you know the convention, you can, of course, change the display to give you the convention you are accustomed to. Or if you are in a situation in which other people control the display, then you can adjust your thinking to the (to you) reversed polarity.

WAITING ON THE INTERACTION

Computers are fast. Your system will probably be fast enough for you at first. You will get accustomed to the quick service, so you'll come to expect it. It will be fastest when it is new, but then you'll add capabilities to it, and more workstations if it's not standalone. It will slow down a little. You'll bear with the slowness for some time, but eventually, either to speed the operations or for some other reason, you will get a more powerful computer. Displays will whiz onto the screen. With that power, there will be more things you can do. You'll add power-hungry programs that do wonderful things. You'll use more interpolated data. More people will see the advantages of using the system. Everybody will be happy with it. Still more programs and still more people will be put on it. It will start running slower. . . .

What you will notice most is that displays are slower to form on the screen. The time it takes to display a section depends on the program you are using, the power of the computer, and the load other users are putting on it. If you share a computer with other users—either people on other interactive interpretation workstations or people in several other departments of the company—then when use is heavy, you are time-sharing, with the time going to more users and/or to users of more power than normal. Your work slows down. Don't feel put upon, though; theirs is slowed too.

This is just a way of taking turns. There is another way that time-sharing may be affected, though, by assigning priorities. Naturally, an important task or one that must be completed by a certain time should get more of the computer's attention than one that is less vital or that doesn't have any special time limits. So the computer can be set to do the time-sharing so the high-priority tasks are slowed less than other ones. If your work happens to have a low priority, you will find yourself waiting longer than if it didn't.

And sometimes a computer slows down for unknown reasons, when maybe only one person is using it for an ordinary interpreting task. And if the computer is doing some kind of task in batch mode, it may compete with your interpretation, so your activities are slowed.

Waiting on a computer while it performs a task can be a nuisance or not, depending largely on your attitude. The wait is usually a few seconds, almost never nearly as long as it would have taken you to do the task by hand. Doing the work by hand, you wouldn't have been impatient, because you wouldn't have been waiting, you'd have been busy. Instead of feeling impatient, you can let the short waits on the computer be a time to let your eyes go unfocussed to rest them, or to think about what you are working on.

When you display a section, the section forms at moderate speed, starting at one edge and taking shape across the screen. While that is happening, there isn't a cursor on the screen, so you can't do anything. And when the section is completely formed, if you have told it to display some picks, then there still isn't a cursor.

You continue to sit there while the horizons and faults that you've picked on that line are drawn, and while the symbols for picks on intersecting lines are drawn. Any wells that are close enough to the line (within a default distance or a distance that you have set) must also be drawn. Only then does a cursor appear so you can interact with the interactive system again. But if you decide you want to see the display at a larger scale, and give a command to zoom it, then you have to wait again while the computer forms the smaller part of the section that will appear at the larger scale. And you wait while it draws the picks again, and the intersecting picks, and the wells. The smaller amount of the section has less of all this information, so it is all plotted faster than the whole section was, but it still takes time. And if you decide to go to a smaller scale to show more of the section, the re-drawing takes longer because there is more information for the computer to handle.

A system may have an interrupt function that lets you stop any display you don't want to see completed, so you can immediately go on to the next command.

Some systems do use the temporary storage described earlier as a basis for their displays, showing the data in that storage as small displays of sections or slices on the screen in columns and rows. Any part of the temporarily stored data can be displayed quickly as an enlargement of part of the whole set.

On some systems, a composite line, made up of parts of different 2-D lines or of lines created in any directions from 3-D data, can be stored so you don't have to put it together again if you want to use it later. For instance, a line through several wells might be one you would want to refer to occasionally in the process of interpreting an area. That saves time not only by your not having to point to the parts to put it together again, but also by the computer's having it stored as a unit, like any other line, so it doesn't have to hunt up the different parts of the line.

What else can you do while you are waiting for a display to appear on the screen? It depends on how pressed for time you are. If there is some other activity you want to do, one that can be picked up and dropped easily, and if you are not in a hurry with your interpretation, then you might be able to sandwich them together. For instance, while you are new to the system, you might want to read parts of its manual between giving commands. Each of the two activities would have to wait while you got to a stopping place in the other, but your time would be used fully. Depending on the urgency of the interpretation, dividing your time can be a fairly good idea or an extremely bad one. The decision also depends on where you are in the interpretation. If you are trying to solve a difficult problem, you might not want your thoughts interrupted while you are concentrating on the problem. In some other solution of a problem, the waiting time might be just what your mind needs to seek out solutions to problems in interpreting.

SAVING TIME BY DISPLAYING LESS

In general, you'd rather not spend much time waiting on displays to appear on the screen. One way to speed the displays is to not have the computer do so much work to show one of them. Cut down the area displayed; if you are interested in the east end of a line, don't display the west part. The computer has to call up each trace from memory, so you can save time by giving it fewer traces to call up. Display a short range of times, so the computer doesn't have to call up so many samples for each trace. One way to do both is to zoom the section up to a large scale. The screen doesn't hold many traces at that scale, or much length of a trace, so it can call them up quicker. The system also takes time to display picks, so you can speed the display operation by having it display the picks of only the horizons and faults you really need to see at the time, instead of just regularly having it show all the picks you've made.

If you have one section on the screen, and are going to call up another and then enlarge the new one, you can save yourself some time. First enlarge the scale of the section on the screen, and then go to the other section. If you had gone to it first, you would have had to wait while the computer displayed all the samples of the section, and then zoom up. If you zoom up first, then only zoomed sections, with less time range and fewer traces, are displayed.

Some fancier displays may take longer. If your system doesn't have enough computer power for all the use it gets, it may take so long to generate attribute displays that it isn't practical to make them as you need them. If you have plenty of space on hard disks, you can have the system make the displays at night as batch jobs, so you can call them up at any time. But if you also don't have much disk space to store them on, you may have to make some sort of compromise, maybe making a few in batch mode, using them, and then deleting them. It may be faster and more efficient to do most of your interpreting on simpler displays, and just use the others for solving occasional problems.

When you get impatient waiting, just remember how long it takes to interpret on paper. And you may remember a time when you spent half a day just hunting for the paper sections to work on. DOING SOMETHING ON A COMPUTER IS SO MUCH FASTER THAN DOING IT THE OLD WAY, THAT YOU CAN'T STAND TO WAIT ON THE COMPUTER A MOMENT.

COMPUTER PROBLEMS

Expect some things to go wrong.

When you work with a computer, there are special computer problems you have to contend with. They aren't particularly worse than other kinds of problems you've dealt with before, but they are different. I'll start with a personal experience.

It was pleasant working on an interactive system one weekend. There weren't any other people wanting to use the equipment. I was alone with the system. It was a good time to get work done. The only problem was that the office was hot. The air conditioning was off for the weekend. I was hot and uncomfortable. A fan blowing across me helped a little, but the room was still pretty hot. That wasn't a great handicap, though. People can work quite efficiently under miserable conditions. Enthusiasm for the project, a frantic deadline, or some other driving force can prevent you from even paying much attention to physical discomfort.

So I was interpreting away, sweating but enjoying a quiet time with the system, getting pretty much done. I had a section on the screen, displayed in some colorful combination like red, blue, and some intermediate color. Then a strange thing happened. The section changed color, becoming mostly yellow. I wondered what was going on. The yellow phase didn't last long though, and my bright colors came back. There wasn't anyone that I could ask about the problem, so I kept interpreting. Some time later, the screen went yellow again. There were a few episodes of this computer jaundice, but I was generally able to keep going. Then the yellow came again, changed to a paler yellow, and the screen went black. That put an end to the session. I shut everything down and went home.

On Monday, the system worked fine. I reported the problem, and the technical people looked the system over but couldn't find anything wrong. Well, you can guess what the trouble turned out to be. I could stand the heat, but the computer couldn't. Or, to be fair, I could stand the heat out in the room, with a fan, but the computer couldn't stand the higher temperatures its electronics generated inside its case. Getting the air conditioning on over the weekends turned out to be a difficult communications problem. But there was a solution to the overheating. The technician took the back off the computer, and I put the fan so it blew back there instead of at me. The fan blew that extra heat out into the room, so it couldn't build up inside. It was hotter than it had been where I was, but the screen didn't go yellow any more.

A simpler problem is just having the image get a little fuzzy. There is a trick to make it sharp again: turn the screen's power switch off and on. In coming on, it automatically clears up any stray magnetism that may have been picked up. That may make the picture sharp. If the improvement lasts until you finish the day's session with the system, then you don't have a problem, because when you turn the screen on the next day, it will automatically demagnetize again.

Earlier in the book I discussed commands as ways of telling the computer what you want it to do. But you will be shocked to find that it sometimes doesn't do what you wanted at all. It may do some strange thing, and you may not even be able to tell what it has done. Almost always, the trouble is a "finger problem"—your fingers giving the wrong command. You either momentarily forget which command to give, or accidentally press the wrong button on the cursor-mover, or make some other simple error. Sometimes you will press the correct button or select the correct words or icon, but what you expected to happen doesn't occur. It may be that you didn't wait until the preceding action was completed, or in the case of pressing the button, you may not have the cursor within the appropriate area on the screen: within the seismic section, on the scroll bar, or on a menu choice.

These problems are worst when you are a beginner with the system, but some of them may increase when you gain skill and then become overconfident. They can give you a panicky feeling when you aren't sure how to get out of the situation they put you in, or they may just annoy you when you know what is going on.

A problem you will have with the computer quite often will be from not knowing what command to give to get some result. Sometimes you won't know there is a need for a command. You might tell the system to calculate an attribute and find that the display you get isn't of the area you had on the screen. Why not? You try again, and again get the same part of a section. If you're not thinking sharply that day, that problem can leave you totally incapable of making the displays you want. If you then ask someone for help, you may be told that there is a special command for giving the extent, in both reflection time and shot points, of the area to be calculated. How simple and reasonable, but how elusive if you are inclined to assume that it performs its function on the part of the section that is on the screen!

Or maybe you are working with a function that really is performed on the area on the screen, but forget that you have zoomed in to enlarge a small part of that area. So your function is performed on just the area shown in the zoomed display. Later you call up the results for a number of lines, and find that, starting with the line you had zoomed in on, they abruptly change and are made only for that smaller part of each section.

There are a number of functions in a system that can't be stopped once they are started. If you give the wrong command for whatever reason, you may be stuck with waiting until the process you didn't mean to start is complete. Then, if necessary, you can undo that process and start the one you wanted. After having that experience a few times because of one particular command you tend to give incorrectly, you have incentive to learn to give it properly. There are special commands to abort some specific functions, and on some systems some processes can be stopped with "Control C", that is, turning to the keyboard, pressing the key marked Control or CTRL and holding it down while you strike the C key. Control C is a general computer command that stops many activities. As such, it might possibly take you out of the interpretation program, though. Learn what it does on your system, by reading or by experimenting, before depending on it.

The wrong-command problem, as I said, accounts for nearly all the strange things the computer does. In general though, you can't harm a computer or a program by hitting the wrong keys while using the program. But you can do some things that you won't like. You can overload the disk memory so the system refuses to accept any more, or even loses what it has. You can cause a system to "freeze up", so it won't accept commands. Then it may take a specialist to get it going again, or just someone who has had that particular problem before, and can act superior while getting you out of the mess. Your problem may be as simple as not having something plugged in or switched on.

You will sometimes have to phone a vendor about a problem. That may call for one of the vendor's technical people to go to your office and try to eliminate the problem from your interactive interpretation system. Some problems may not require a visit, but instead may be solved over the telephone. You may be "walked through" a diagnostic process. A specialist will talk to you, asking you what happened, then having you give certain commands and report on the results. You are providing the fingers and eyes for the expert to operate your computer. That is pretty effective, and often solves the problem. The walk-through by telephone can be the only way to solve a problem quickly at night, or if you and the vendor are in far-apart locations. A walk-through can also be used to supplement training. In that case, you don't know how to do something, and can't work it out from the manual. So you call the vendor, and someone leads you through the steps you need to take. After going through it successfully, you probably should ask the person to describe the steps again while you take notes. Doing it correctly once doesn't mean that you will necessarily remember later.

There are problems you can have with computer peripherals: plotters, printers, disk drives, tape drives, etc. These devices are mechanical, with moving parts that have to be adjusted to close tolerances.

A plotter can run out of paper, ink, or ribbon at inopportune times, like when you are in a rush to get some displays ready for a meeting, or when you are alone in the office at night and either don't have access to the supply room or don't know how to load the machine.

A plotter might make perfectly good hardcopy, except that your display is cut in half, with the top half printed below the bottom half, so it looks like a television screen with the vertical hold out of adjustment. A plotter can have other problems that mechanical devices are subject to. It is often best not to try to get a plotter working yourself, but to call a technician who knows how to work on it without causing new problems. If you must have printouts right then, maybe you can switch to another plotter.

The reason for my stressing the word "mechanical" is because computers are not mechanical and so don't have those problems. The main working parts of computers are electronic. Their work is done by electrons moving around in and between chips. No ratchets move, no wheels turn, no parts are worn by friction. So the computer has far fewer problems than mechanical devices have.

That "no moving parts" is just the interior of the computer itself, of course. The same computer uses disk drives, and they are mechanical. But they are fairly simple—just fast-spinning disks with magnetic heads, like a tape recorder has, to put information on them and read it from them. The disk drives are expensive and well-made, so they are reliable. However, the possibility of a "head crash" of a disk is one of the reasons for making frequent backups of interpretations. A head crash—the drive head actually plowing into the surface of the disk where the magnetic information is stored—is highly unlikely nowadays. Although people still worry about hard-disk crashes, these modern "crashes" are usually not hardware problems at all but caused by flaws in programming or by human error. Information may be wiped out of memory, or "garbage" (random data) put on the disk in place of some of the information.

When the computer does do some crazy thing, like putting garbage into your work, you may be able to straighten it out by going over the interpretation and correcting the errors that were introduced. But a much smoother way of repairing the damage is to have a recent backup, get the interpretation from it, and re-do the work you had done since the backup was made.

There are problems that are caused by "bugs", flaws in the program. A program for seismic interpretation is large and complex. It must translate a simple command you give into a series of detailed commands to the computer, to a series of ones and zeros for the computer's binary nature. Different commands and functions may affect one another. Writing a program is a long, tedious, complex operation. A lot of the programmers' time is spent on checking what they have done. But the checking can't catch everything; some of these bugs will remain to be discovered by users. A programmer will work out a function, but an interpreter, many interpreters, will use it over and over. It is probably correct to say that all large programs have bugs in them, some more than others, of course. Newer programs naturally have more bugs than older programs, because users report problems and programmers fix them.

What do you do when you run up against a bug in a program? The first step is to make sure it is a bug, and not the more likely wrong command you gave. To check that, give the command several times, being very careful to do it correctly. Get out the manual and make sure you know how the command should be given. Ask other people who use the program, you may be consistently overlooking something. If you do determine that it is a bug, inform someone who can do something about getting it corrected. That may be a programmer, your boss, the manufacturer, or a manager of computer systems in your company. Unless and until it is corrected, try to find a way around it by using some other command that accomplishes the same thing or nearly enough the same thing for your purposes. Or maybe use the flawed command and then follow it with a command to undo the damage.

What if some of your interpretation is lost? It would be frustrating to put a lot of time and effort into an interpretation, and suddenly find that all you had done had vanished out of memory. If that sounds scary, it's meant to.

There is no question of whether you will sometime do something wrong that loses some of your interpretation—you will. And worse, you will cause someone else's interpretation to be lost. And sometime or other, another person will happen to lose your work, maybe by deleting the wrong thing in making space on a hard disk. And sometime the equipment will lose your work. So it is vital to have a way of keeping these losses from being catastrophic. Making frequent backups is the answer.

BACK IT UP!

In normal living we learn to be careful with information on paper. We have a quite complicated set of knowledge about how to keep from losing data that is on paper, by keeping papers in notebooks or file cabinets, not letting them get wet or scorched, or something. The interpretation stored on disk is also subject to loss, but in different ways. It is easy to delete something by mistake. Or sometimes a machine breakdown or a power failure can cause you to lose some of your picks.

The basic uninterpreted data is pretty safe from loss of this kind, because the original of it is still on tape somewhere. And the interpretation system normally doesn't let you change the basic data it has on disk. It is stored as "read only" data. But the results of your interpretation are "read and write" data, that can be erased and recorded over. They have to be, so you can change your mind when you need to, and keep refining the interpretation. This interpretation information is subject to loss until it is complete and stored in some fairly permanent way.

The interpretation data is probably stored on hard disk automatically as you work on it, in a way that doesn't interrupt your work. But that is just one copy. It is usually safe, but is vulnerable to all sorts of rare calamities that can lose part or all of it.

How do you keep the losses small? By making copies of the information frequently. These backup copies preserve the information as of the time the latest one was made. If you back up your interpretation once a month, then you could lose nearly a month's work some time. If you do it once a day, then you can't lose more than a day's effort. Imagine the effect on your career of working against a deadline—lease sale, lease expiration, rig move, or something—having the interpretation almost done just under the wire, and then losing it all at the last minute. Still being scary? Sure. I'd rather scare you than leave you with false confidence. Backing up is some trouble, and it takes time that could be used in interpreting, so it's tempting to put it off and get more work done. Just remember that possibility of a last-minute loss.

The usual way of backing up is by making copies on magnetic tape, the same kind of half-inch tape used for recording and processing, although probably on smaller reels. There is a procedure for loading a tape in the computer and instructing the system to copy the interpretation onto the tape. You stop everything, maybe at the end of a day, and go through the ritual. Then take the tape and put it in a safe place AWAY from the system.

Having saved on one tape, the next time you save the information, don't save it on the same tape. Get a fresh tape, and put the new version on it. Take it away too. The second tape serves several purposes. For one thing, you make it without having to bring the first backup to where the system is at the time (the idea is to keep them far enough apart that a problem at one place—fire, flood, vandalism, or a computer breakdown—even during the time you are backing up data, won't affect the other copy). For another, in case part of the new interpretation you did turns out to be a mistake, the older version is still available on the first tape.

Then, the third time you make a backup, get a third tape. The fourth time, you can bring back the first tape and replace the interpretation on it with the newest version. This is called the "grandfather-father-son" method of backing up.

Keep the three tapes circulating, so there are always recent versions of the interpretation safely stored away on tape. This routine is laborious, but when one of your interpretations does somehow get lost, as it certainly will, the cold sinking feeling that comes over you can be replaced by a tremendous feeling of relief when you realize that you do have most of your work saved.

The backups the system may make automatically on hard disk are fine, but you don't dare depend on them. The hard disk is part of the computer setup, and stays in the machine. You also need frequent backups made on tape, so they can be removed and stored in another location.

Some backups are made on videotape, regular VHS or 8 mm videotape. The tapes are readily available because they are made and distributed for home videotaping. Both the tape and its recorder are small and inexpensive. This arrangement makes backups much faster than regular seismic tape does, but the videotape doesn't last through re-recording as well as standard seismic tape. So the videotape is discarded after being used several times. Even so, it is much cheaper than half-inch tape. It is considerably more compact, especially in the 8 mm format. Being in a cassette makes it easier to handle than reels of tape.

In a large organization there is probably a regular system of someone's making backups of everything on tape at frequent intervals, perhaps every night. In a small company, or one without a computer department, you may need to make the backups on tape yourself. Just make sure it's done, one way or another. BACK IT UP, KEEP BACKING IT UP, BE SURE THAT IT'S BACKED UP!

SAFEKEEPING AND SECURITY

One great advantage of interactive systems is in the simple legwork of filing data. Interactive systems take care of the filing and handling that are problems in working with paper sections. For example, any part of any section can be displayed just by giving the commands that call it up. It isn't misfiled or lost in the heap on your desk, and it isn't squirreled away in somebody else's office.

But you do have to keep your tapes, and your optical disks if your system uses them, labeled and stored in their proper places. They too can be misplaced, but it doesn't take much effort to keep them where they can be found. Data on tape or disk is many times as compact as data on paper, and isn't on a lot of separate pieces, so orderly storing is relatively easy. But if you do somehow manage to do the unthinkable and misplace a tape or a disk, that is a lot of data lost. So backup copies in another place are essential.

In interpreting with paper sections, geophysicists have formed habits about the security of the data. Sections are kept in a reasonably safe place, at least in an office that is locked at night, often in locked cabinets. Contour maps have information that is more directly useable in leasing land, so they are kept even more carefully.

Now here come these interactive systems that hold all the sections of an area on a single disk, maybe a removable optical disk. And at the end of the interpretation the mapping of the interpreted picks may be put on the same disk. That disk is VERY VALUABLE. It needs all the security of the sections AND the maps. And it is less bulky than the paper sections and maps of an area. You don't want to be so careful that you can't do your own work effectively, but you should be very aware of its value, and take proper care of it.

And if your company makes backups on videotape, the compactness of the tapes makes the problem more severe: a cassette is smaller than a reel of tape, so it is easier to misplace or be taken away. Your company needs to have special arrangements for keeping them in a secure place, and keeping good track of them.

In one office we had an ideal place to store data. That was before interactive systems, so it was data on paper, but it would work for both. The space our company had rented for the office happened to be equipped with a bank vault. The door was steel and more than a foot thick. It was opened by a combination that was entrusted to only a few people. Those who had the combination soon learned to open the door carefully. It was so heavy that a considerable pull was needed to start it opening, and when it got started, the momentum of all that mass kept it moving slowly but forcefully. It wasn't wise to be between that massive door and the wall while the door was inexorably opening. The vault was a great place for secure storage, although it was in the same building as the office.

Where do you put the backups away from the office? If your company has several offices in different places, they could store one another's backup tapes. But that means that to restore lost information, you'd have to wait for, and depend on, some means of delivery between offices.

A natural place for storing backups is a tape-storage service. These storage places are set up to handle large volumes of tape. Your company may have many square feet of space at one of these air-conditioned warehouses filled with the field tapes and final unfiltered tapes of all of its shooting. There is plenty of room for a few extra tapes of interpretation.

These storage services usually have a pickup-and-delivery service, so you can call them to pick up a new backup tape and to bring you one that you are ready to re-use. There is a fee for that, and the service isn't necessarily quick, but neither of those is a problem. The fee is tiny compared with the risk of lost data and, since you will probably be using the grandfather-father-son system, so they already have fairly recent tapes stored, there is less urgency to get the tapes to them immediately. If you need a tape in a hurry, to retrieve some of your interpretation, then no doubt a quick delivery can be arranged.

One good answer is a bit like the bank vault we had. With the extreme compactness of some forms of backup, a convenient place for storing backup interpretation data is a safe-deposit box at a bank. These boxes are not expensive; they are in secure vaults; and the vaults are air conditioned to a temperature and humidity that are good for magnetic tape. You can get a box large enough for at least small-diameter reels of tape. Or if you use a videotape backup system, the cassette will fit easily in even some of the smallest safe-deposit boxes. That storage may also be convenient, as there is likely to be a suitable bank near your office.

4

Interaction

One of the big advantages of an interactive system is that it adds to the fun of interpreting. Anything that does that for your work makes your life more pleasant and lets you do better work (Fig. 4–1).

The list of abilities of an interactive seismic system is wonderful. You can find it hard to tear yourself away from the system. And with the enthusiasm and the convenience, you may well be stimulated to think up more ideas. Certainly you'll be more inclined to test ideas since it won't take long to try them. You may not realize how many ideas you had put aside in the past because it would take too long to try them out.

LOOK IT OVER

There is a piece of advice you hear occasionally about reading a technical book or article. The advice is to skim the whole thing first, so you get an idea of what is covered and what points it is leading up to,

Geco Geophysical Company, Inc.

Fig. 4–1 Workstation

and then to read it. That may or may not be good for its stated purpose, but a similar technique is good for doing seismic interpretations.

The method is to look over the data before getting into the interpretation. "Waste" some time on the data. Fiddle around with different displays and different color schemes. Time spent this way will be surprisingly useful to you when you get into the details of the interpretation. You will have some idea of the overall reflection quality, major structural features and faults and the kinds of displays that bring out characteristics better. That once-over-lightly is useful when working on paper, but interactive interpretation is especially good for making it easy to get such an overall look at an area.

One of the advantages of interactive interpretation is its ability to show sections and slices in a sort of movie form, one image after another fairly rapidly. It isn't rapid enough to give the impression of motion the way a real movie does. The expression "paging through" is used to describe this slower changing. You have to wait, maybe a few seconds, for each display to take shape, usually starting at the left edge and building to the right. The quickest way to get an overall view of a 3-D area is to watch it in this movie form. Look at a movie of time slices, then a movie of sections.

If the area is 2-D, you can still look at the sections in succession, although one line won't look as much like the next as in a 3-D area. In either case you get an idea of the area and its major features. And if you are new to the system, this leisurely looking can help familiarize you with the commands. Then you can view the area more slowly by displaying a few of the sections one after another and just perusing them.

Call up the base map on the screen and give it a looking over. If there are wells in the area, check to see if they are all there and plotted correctly. This could save you much work later, trying to tie a well that is mis-spotted, or keep you from recommending drilling at a location that has already been drilled. Checking pays off. Even if you don't find any mistakes, it makes you more familiar with the area.

Take your time looking the area over. It may be the best time you spend on the interpretation. The alternative, getting down to work immediately in one place and extending your interpretation bit by bit over the area, is a gross waste of time. It can lead to your being surprised by some feature, and having to start over, maybe after having worked a large part of the area. You might be working a reflection, only to find that it degenerates so much in part of the area that you can't pick it there, and have to change to another horizon.

After inspection of the data, you should have a fairly good idea of what to expect in the area as a whole.

CHECK IT OVER

As in all work with quantities of data, you have many chances to make mistakes. Interactive interpretation frees you from a lot of checking. In working on paper you have many chances to make mistakes. You can read the reflection time wrong on any shot point. You or whoever else puts the times on a map can plot any of them incorrectly. In interactive interpretation all of that is done for you, quickly and correctly. You don't have to worry about those tasks. The computer will do them right almost every time, and when it does do something wrong, it will probably be so wildly wrong that you will notice it immediately.

But there are still many chances to make mistakes. And the greater quantity of data you can handle increases the opportunities for error. Computers make people so powerful that they can make bigger mistakes. What can you do about the possibility of mistakes, be careful? Of course you should be careful, but that isn't good enough. No matter how careful you are, there will be times when you just do something wrong. It comes with being a person, and no amount of care or experience will immunize you against it.

Worse yet, the error that slips by your defenses may be the very one that someone else, with attention on different aspects of the work, notices at a glance and makes you feel foolish. Or worst of all, the error may not be noticed by anyone, and may go on to bring about a dry hole.

So during the interpretation, what do you check? Just about anything you can check. Especially if something looks a little strange, don't take it for granted that it is correct. If something seems strange, check whatever you can that may bear on it—well locations, seismic line numbers, even the area name.

There isn't any amount of computer power or even expert advice that can keep people from making mistakes, so watch out.

COMPUTER PICKING

Picking on an interactive system is fast and convenient (Fig. 4–2). There are several degrees of automation of picking available on an interactive system. A simple way of selecting one or another of them is provided, by punching a button or by moving the cursor to a command on the screen and clicking. Like other features, these different ways of picking may not all be on a particular system.

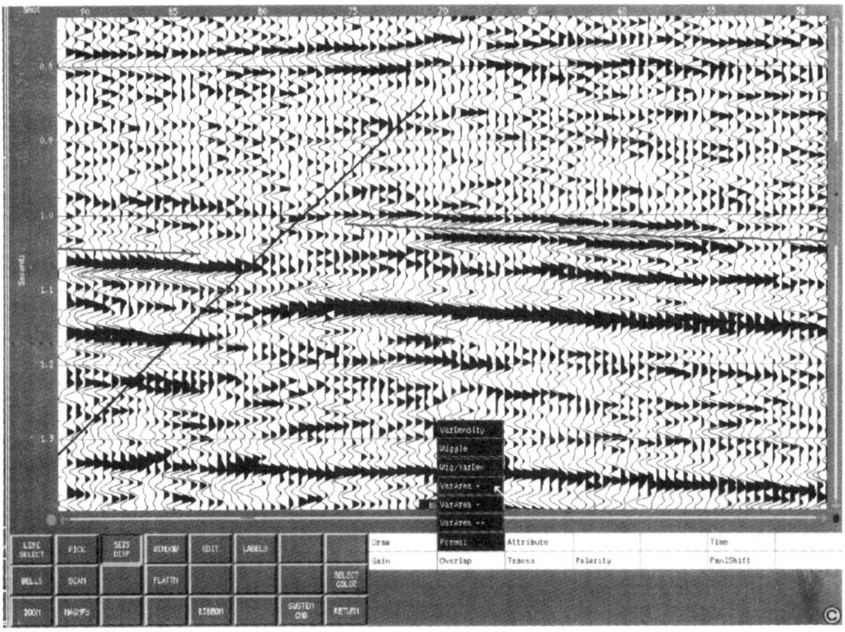

Sierra Geophysics, Inc.

Fig. 4–2 Picking on screen

FREEHAND

The way of picking that is the most like marking a paper section is moving a cursor-mover on the desk, with the cursor on the screen imitating the movements of the device and leaving a colored trail behind it like the line left by a pencil. Although this method is pretty much like marking on paper, there are several differences. For one thing, in order to follow a reflection, you have to move the device on the desk while watching where the cursor goes on the screen. You get used to it quickly. It's as natural as turning a steering wheel to make a car go where you want it to. It has two big advantages over marking on paper: your hand doesn't block your view, and neither does its shadow. There is also a disadvantage. If your hand shakes a bit when you move the cursor-mover in this freehand way of picking, the line you make is irregular. Drawing on paper, if your hand shakes, you mark over the pick a few times to cover the irregularity so it isn't noticeable. But on the screen, you have to delete part of the pick, and then re-pick it. It may seem to go too fast for you, making you wish for a more deliberate way of picking. There is such a way, using straight lines.

RUBBER BAND

To use the straight-line method, move the cursor to the point on the reflection where you want to start picking, press a button on the cursor-mover to anchor that point, and then take the cursor to another point on the reflection. The program is using its "rubber band" mode, in which a visible line stretches from the chosen starting point to the cursor, like a rubber band stretching between starting point and cursor. Wherever you move the cursor, that rubber band clings to it, always connecting it to the starting point. Move the cursor somewhere farther along the reflection, and click a different button. That sets the line so it stays there. Move the cursor again, and a new rubber band follows you, anchored at the last place you clicked. Move to another point and click again. This straight-line picking works fastest where the reflection is fairly straight (either flat or dipping, but not curved), because you can follow a reflection with long segments of line. If the reflection is not straight, it's still easy to use the rubber-band method, but you have to pick in many short segments.

Both these methods, drawing the pick as you go and picking by rubber band, are manual.

SNAPPING AND SMOOTHING

Picking in partly manual, partly automatic mode may usually be the most satisfactory. Before starting, specify what part of a reflection the pick is to be made on, that is, peak, trough, zero-crossing after a peak, or zero-crossing after a trough. Then pick just the way you do in one of the manual ways, freehand or straight-line. Then give a command, probably by clicking, and the computer moves the picks up or down to the nearest trough or whatever part of the reflection you had chosen. The pick "snaps" over to the event. You've done the picking roughly, and then the computer makes the picks more exact.

You can set the system to snap automatically. Then snapping will occur every time you indicate that you are at the end of a segment of continuous picking, for instance at a fault or other break. If you don't indicate a break, then when you leave picking that horizon on that section (by making a different horizon active or by displaying a different section), what you have picked will be snapped. There is a disadvantage to this automatic snapping. Some parts of the pick may wind up being snapped several times. Each of those times takes a while to do, so you spend more time waiting for the computer to do them.

You will probably have to ask the computer to snap a segment of line by pointing to the segment. There may be a command to snap the entire area, not just what is on screen. But it is usually safer for the snapping to occur only when you are watching, so you can bring the pick back on track if it goes astray.

Snapping makes a pick irregular. The highest amplitudes of an event on adjacent traces jump around, one high, one low, etc. Picking on paper you would mark a smooth line through those little irregularities. The system has a way to handle it too, a command to smooth the snapped picks. It runs a moving-average type of calculation, so the ups and downs are made more gentle. Like snapping, smoothing usually operates on a continuously picked part of a reflection, and works only on the section that you have on the screen at the moment. If the smoothing doesn't take out enough of the irregularities, you can smooth again. That smooths the smoothed pick some more. You can have as many stages of smoothing as you want.

The functions that snap and smooth take computer time. Whether that delays you noticeably or not depends on the power of the computer that runs your system, and also on how much of that power is taken up in time-sharing with other users or processes. If, either because your computer isn't very powerful or because of time-sharing, snapping and smoothing slow your interpretation more than you want, there are a couple of tricks that can reduce the time spent in those activities.

The first trick keeps you from snapping and smoothing the same pick several times.

Try picking a horizon over the whole area by rubber band, without snapping or smoothing at all, and then, when it is all picked to your satisfaction, go back through the lines and snap and smooth them. You save time by not polishing picks and then later changing them and then having to polish them again. But you spend extra time calling up every section again when you do the polishing. Whether it gains time for you or loses it depends on how much you have to change picks (probably more than you thought beforehand) and how long the polishing takes at the end.

The second trick keeps you from having to polish in little bits.

If your system does snapping and smoothing on segments of picks when you tell it to, then in a faulted area or an area with reflections in pieces separated by breaks, you may spend a lot of time telling it to smooth this segment, then that one, etc. One way to simplify that is to pick the reflection as one continuous pick, bridging across the breaks, then snap and/or smooth, and after that delete the bridges. Snapping and smoothing are each caused by one command for the length of the pick on the screen. It will take less of your time, and also less of the computer's, since it won't have to get started several times. After doing that, though, be sure you don't forget to **break the bridges.** Don't even put it off, but do it as part of the operation, so there isn't any chance of your forgetting to.

In 3-D areas, time slices present a special problem in picking. The reflections look wide on slices, so it isn't easy to pick precisely on them. Your system may let you pick roughly through a wide event, and then snap to the exact positions. The snapped picks look strange, jumping back and forth within the wide event. However, when those erratic-looking picks are displayed on sections, they are on the correct positions of the reflections, which appear narrow on the sections.

AUTOMATIC PICKING

When the reflection is good enough, the easiest way to handle the picking is to have the computer do it for you. This is automatic picking. It is based on comparing one trace with the next. There are several kinds of logic used in automatic picking. Typically a program compares adjacent traces, looking for a peak or trough on one trace within a range of angles from the time of the same event on the preceding trace. The angle is based on the expected dip in the reflection. The angle is not the subsurface dip angle, but an angle between the event on adjacent traces, largely determined by the amount of irregularity from trace to trace. If you find that there is steeper dip than you allowed for, or that you allowed for too much, you can change the specification and re-pick.

Or, instead of dip, the system may let you specify a "window" (not the same as the window that is a segment of a section or map on part of the screen) of so many milliseconds above and below the time of the event on one trace within which it searches for a similar feature on the next trace. With a wide window, a good reflection can sometimes be picked fairly well in automatic mode, even if there are small faults or other breaks in the continuity. But if the reflection is poor, you need a narrower window. If the reflection of that horizon is high-frequency, you need a narrow window because a wide one would include other peaks or troughs of the reflection, and the automatic picking could jump to one of them. If the reflections dip steeply, you need a wide window to include the time of the reflection on the next trace. The best way to decide what size window to use is to try different ones on a sample of the data, and see which is most helpful to you.

To do automatic picking, mark a starting point on the reflection, then indicate how far you want the program to pick. Put the cursor that distance away on the section, at any seismic time, not necessarily on the reflection. The computer decides where the reflection is all along the distance you specify. Alternatively, on some systems, you indicate a starting point and some other points on the reflection, and the system automatically picks from the start to the next point, then to the next after it, and so on. That is a little like rubber-band picking, but with the line not visible. You don't need to snap it to the reflection, it is already snapped. That is, the automatic picking looks for the exact points, the peaks or troughs or zero-crossing points, the same ones snapping finds. You can smooth automatically picked reflections the same way you do snapped picks, using the same commands to smooth.

One strange-appearing feature of automatic picking (or snapping) occurs when you are using it on a wiggle-trace display. It's easy to see that in picking a zero-crossing it goes right to it. But automatic picking on a peak or a trough doesn't seem to fall on the apex of the wiggle, but to miss it a little. Is the automatic function so sloppy? No, it is accurate; the way people usually pick on paper is sloppy. A peak or trough shows information about the amplitude of the trace, and the central position of the trace shows where the trace is located. Peaks and troughs swing to the right or left of the trace position to show what the amplitude is. The system goes to the central position of the trace, and to the seismic time of the peak or trough (Fig. 4–3). If a peak or trough from another trace happens to reach that straight-trace position, it isn't necessarily at the same reflection time, so the pick seems to miss it. If you doubt a pick, just compare its time with that of the peak or trough for its trace. On a variable-density display, the automatic picks don't look incorrect because the traces don't wander off to the side.

Except on good reflections, snapping tends to be better than automatic picking, because the automatic can easily go astray. It wastes time wandering around, and then you have to delete its wandering, pick manually and snap your pick. After going through that a few times, you may decide that it would have been quicker and would have given you exactly the same accuracy if you had picked and snapped in the first place.

Seismic reflections are usually nearly parallel with the reflections immediately above and below them. They have differing dips at the boundaries

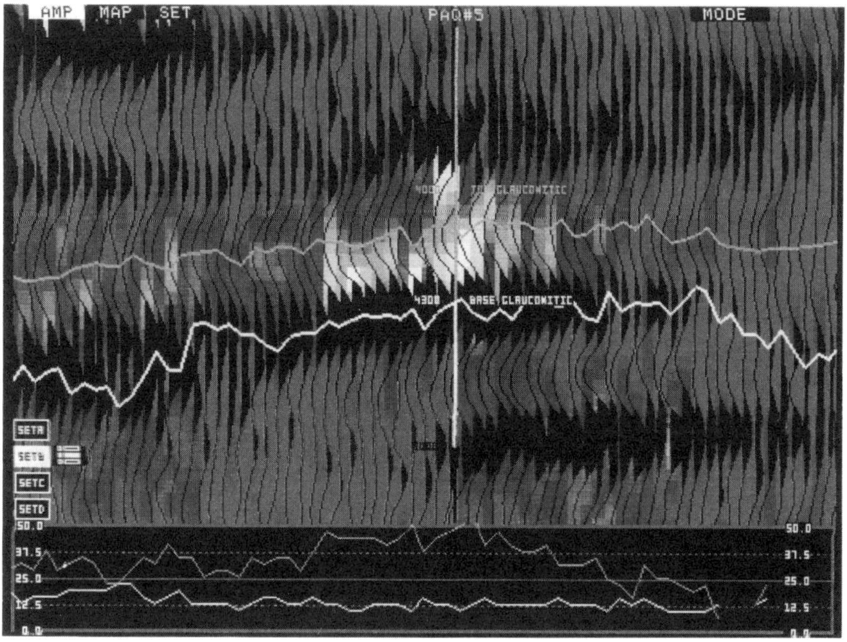

Graphic Science Corp.

Fig. 4–3 Automatic picking on wiggles (see in color on cover)

of seismic sequences, of course, but there are more layers, and therefore more reflections, between those boundaries. And even at the boundaries, the difference in dip is often small. So it is frequently safe to assume that reflections near one another are almost parallel. A glance at the section will usually tell you if they are not. This near-parallelism can be used to help in picking. A system can, after a reflection is picked, snap to another leg. You may pick a peak, save that pick, and then use the snap function but snapping on the trough below the peak. It's just a matter of the snapping parameters you give it. Tell it to snap on a trough, and to look in a range of times below the peak. Since it is snapping, it is picked exactly on the trough, so you don't need to do any adjusting, unless there is an abrupt change between the peak and trough. Carrying this idea further, some systems can, as you pick a reflection, also automatically pick several legs above and below that one. These picks of adjacent legs are useful for analyzing variations in reflections. The time from peak to trough may vary with porosity of the formation, or with some other characteristic. That can be especially powerful in an area with many wells, so you can determine what the small variations mean.

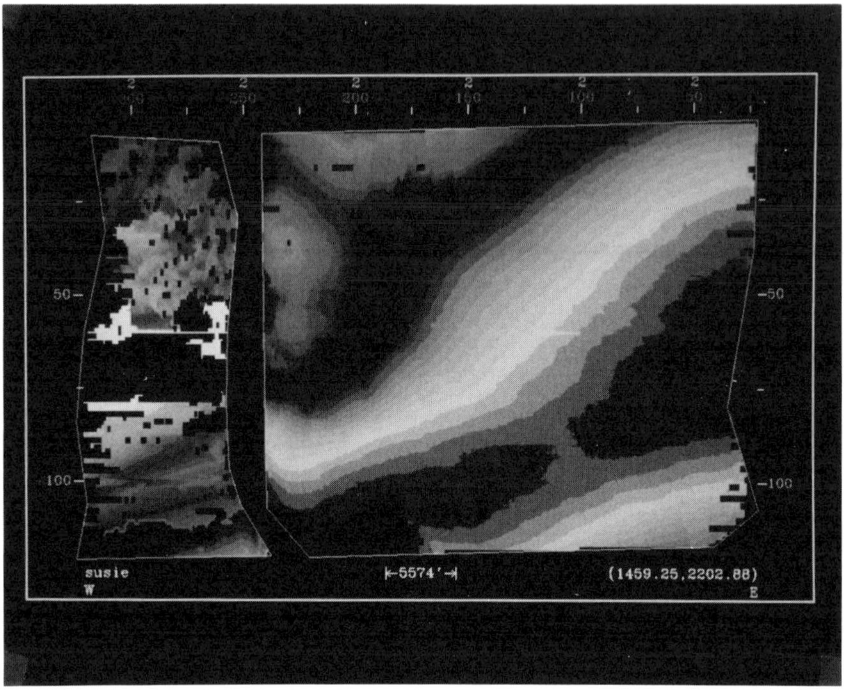

Landmark Graphics Corporation

Fig. 4–4 Automatic picking of area

There is much that automatic picking can do. At times it can pick some horizons all over a 3-D area (Fig. 4–4). Other places it can pick horizons completely in a fault block, and you can then correlate across a fault and give it a start in the next fault block. An automatic system that works outward in all directions from a starting point can pick until it comes to an obstacle, work around the difficult spot, and continue the mapping, so it just leaves a hole where it couldn't pick.

That way, you can rapidly make maps of a large number of horizons. After the horizons are picked, you can have the program calculate the time intervals between them, and map those intervals. The interval maps may help you find features like sand buildups, reefs, buried channels, faults.

ERASING

Erasing a pick is easy, no matter how you made the pick. But some of the ways you erase depend on whether the system allows only one pick for a reflection at a trace, or accepts more picks at a trace.

The simplest way to erase works on both kinds of system. It works while you are doing rubber-band picking. As you pick, mark a starting point by clicking, then stretch the line to the next point, click again, then the next and click, and so on, following the reflection in bits of straight line until you come to a break or the end of the display on the screen. You do the clicking by pressing a button. There is another button that erases within the continuous picking before any kind of break. Press it, and the last straight line segment vanishes, from your sight and from the computer's memory. The rubber band stretches to the end of the segment before it. Press again, and that next-to-last segment is gone. Keep on for as many as you want to delete, as far as the start of that continuity of picking. Then you can start adding line segments again. This is a handy way to pick. Make some straight line picks, decide you don't like the last two, press the Erase button twice, make straight segments the way you think the reflection does go. Do this back and forth until you are satisfied with it.

Other erasing, after you have left a continuity of picking, is simplest if your system just permits one pick per location for a horizon. You can remove a pick by giving a command to put the program into erase mode, then indicating the ends of the part to be erased, and clicking to tell it to execute the command. That length of the pick is gone, as though it had never been. Indicating the ends does not necessarily mean putting the cursor on the reflection at the ends. Like marking the end of automatic picking, it can just as well be anywhere above or below the reflection. It just shows how far the erasing is to extend. Or you don't have to erase at all, but can just re-pick, and the new picks replace the old ones. Since the system allows only one pick on a horizon at each location, it necessarily changes to the new pick when you re-pick it.

For some purposes it is better to have the system allow any number of picks for a reflection on a trace. That is handy for overthrust faults, that do cause the horizon to repeat, in pushing one part of the bed over another. It is also useful if you want to mark a lot of picks on a section, just to see what there is of interest on the section and how easy it is to pick. But with that multiple-pick capability, erasing becomes more difficult. You couldn't very well just indicate the distance to be erased, or all the picks assigned to that horizon would be erased over that distance. So you have to indicate which pick you want to delete as well as how far. If that is done by having you point to both ends of the part to erase, the task becomes tedious. If, for instance, you happened to make a pick divide into two branches, and then decide to delete the upper one, you have to put the cursor exactly at the branching point to keep from deleting either too much or too little. This need for precision can be reduced by enlarging that part of the section, probably enlarging it as much as the system can. Or the system may have a command to delete all of the horizon that was picked continuously, without a break (as for a fault or no-reflection spot). With it you can indicate the upper or lower of the branching picks, and it will just eliminate that one, if there was a break at the branching point. A system may have a function to delete whatever picks are within an outline you draw, so you can clean that part out completely and then re-pick what you want to keep. The same problem exists with other operations that are conducted on a reflection, smoothing and snapping. With only one pick per horizon per trace, those are easy; with more picks allowed, it becomes difficult to specify just what you want changed. But you have the advantages of being able to make more than one pick.

There are a couple of commands that may be included with either one-pick or multi-pick systems. One is a command to delete a pick of a horizon for the length of an entire line. The other deletes a pick for a whole area. That one sounds almost too dangerous to have in a system, but it is useful. You may do some picking that you don't want to keep. It could be just for practice, or to show someone how to pick, or for some temporary use. A temporary use might be to mark a pick on one reflection so you can more easily parallel it on a more difficult reflection. It might be to make an unbroken pick to flatten, even though faults or poor data interrupt the reflection.

A system may have a freehand erase mode in which you just run the cursor along the section, and any picks within some set distance above and below the cursor are deleted as though erased with a wide eraser.

GENERAL

The way interactive interpretation works, you usually pick one horizon at a time. But it is quite easy to change active horizons, so you can pick one for some distance, then switch to another. This is not much different from the way you do on paper, anyway. You can't actually mark two places at once, so you color one horizon a bit and then another, with an eye to how nearly parallel they are and how much they deviate. Interactively you do about the same, but by giving a command to change from one horizon to another, instead of picking up a different-colored pencil. To give the command you may have to select the new horizon from a list, or you may be able to point to it on the section, and click.

Also like in hand picking, you can give each horizon its distinctive color. Once you start a pick with a certain color, the program uses that color whenever you do any picking on that horizon. On a section, you can easily see which pick is shallowest, next shallowest, etc., but on a time slice the special color is particularly helpful in keeping you aware of which horizon a pick indicates.

A section is put on the screen from one part of the computer's memory (or from disk), and the picks from other parts. With the picks thus independent, they can be displayed equally well on another form of the section, on the section at another scale, or even on another section. This layering, with different sets of information stored in memory separately, allows two or more of the sets (layers) to be displayed together, something like having the information drawn on transparent sheets, and overlaying one sheet on another. You can choose which horizons will be displayed on sections or time slices you call up. You may have two or three horizons that you always want to see on sections, or you may want to see only the one you are working on. You can also select which picked faults are to be displayed on the seismic data.

Where another line that you have already picked intersects the one you are picking, the reflection time from the previously picked line is automatically marked on your present line at the intersection. The mark is a small dot, square, or something. The picks that appear are for all the horizons and faults that you have chosen to have displayed. You can see if you've tied or not without calling up the intersecting section.

TO SMOOTH OR NOT TO SMOOTH PICKS

Throughout the history of seismic interpretation, there has been smoothing of data. Interpreters know that the reflection times they pick are not exact. There are assumptions made in correcting for elevation and weathering that may not be quite right. There are little bobbles of reflections, but there probably aren't corresponding bobbles in the rock layers underground. Formations exposed at the surface look pretty smooth.

So in picking on paper, some interpreters have used a straightedge, placing it along the general trend of a reflection and picking straight along it. In contouring by hand, some have made contours smooth by ignoring small differences in the data. Both of these methods work, and have been effective in finding oil. But they can gloss over features that may be important.

With snapping and smoothing, your picks can wind up smoothed. That's nice; it takes out the little bobbles, making your results look more like geology. But in doing so, it has necessarily eradicated some minor differences that might be significant. What can you do? If you don't smooth at all, your results may look pretty irregular. If you oversmooth, they can look geological, but not be very close to reality. O.K., smooth a little, but don't overdo it. That's advice that is easy to give, but hard to implement. How much is not enough? How much is too much? The answers to those questions are matters of judgment. They differ with the area and with what you are looking for. Probably it is a good idea to first look at your data without any smoothing, and then apply just enough to enhance the information you are looking for, but not so much as to blur it. Good luck.

MAPS ON THE SCREEN

If your system has two screens, usually a section is on one screen and a map on the other. If you have only one screen, you can switch the display between section and map whenever you need to. Seismic lines are shown on the map. You can choose how you want them displayed. In a 2-D area, you will probably want to see all the line numbers. If the area is small enough, you may want shot point numbers displayed. In a 3-D area, you will probably want just a selection of the line numbers. A locator can appear on the map like it does on a section, giving the line number, shot point number, and CDP number of the point where the cursor is. So as you move the cursor around, the numbers in the locator change rapidly. Or the locator may appear just when you ask for it.

One way to select a section to be displayed is by pointing to it on the map. You can indicate the two ends of the part you want to display, or you may be able to choose one end and a direction along the line. In either case, if you choose more of the section than the screen will hold, the amount that will fit is displayed, and you can see the rest by moving the display. You can also choose parts of different lines to make a loop or crooked line.

That pointing, though, can be difficult if a large area is displayed, so you have to be very precise to get the right line. The locator helps in that situation; it tells you exactly where you are. But you still have to position the cursor exactly.

To make that easier, or for any other reason like just wanting to see a different amount of the area, you can change the scale of the map. The system may use the same zoom command it uses for sections. Or it may have a special command for the map that lets you choose between two forms of enlarging. One is just zooming in or out, so the numbers, shot points, etc. are enlarged or reduced too. In the other way, the characters and symbols stay the same size while the scale of the map changes, like re-drafting a map at a different scale.

A map on screen can also show reflection times of your picks, and, with some systems, even contours. It can change to serve the purposes of base map, map with data posted, or contoured map. The data and contours are shown for whatever horizon you choose to be the active horizon, ready to do picking on. And, like other screen displays, you can preserve it as hardcopy.

TIMING AND POSTING

When you pick a reflection, the computer immediately determines the reflection time and stores it in memory. If you delete some of the picks, the computer immediately forgets those times. Compare that with timing hand-made picks on paper sections, using a scale to laboriously read each time, and then never being sure you haven't made some mistakes, even after checking the picks.

Another chore in working on paper is posting the picked times on a map, carefully lettering each time at its shot point. In interactive interpretation, the computer assigns the time to the shot point automatically, immediately, and accurately. It will usually not immediately letter the times on a map, mostly because the map on the screen is too small to hold all those numbers and have them large enough to be read. It can letter the times on the screen if you have a map of a small enough area displayed. But that doesn't often happen. The system normally represents times on the screen by a color code.

SHOWING TIMES BY COLOR

To show times on a map by color, one color is used for one range of times, another color for the next times, etc. (Fig. 4–5). You can select the colors for the various times. It is best to not have a great many colors, again because the screen is small. The steeper the dip, the greater the difference from the smallest to the greatest time. One color might represent a span of ten milliseconds, or twenty, or a hundred, depending on how much dip the area has. The system may determine the highest and lowest points on the map, assign the end colors to those two points, and then proportion the others evenly in between. Or it may let you decide how large a range of times each color is to represent. In either case, it is similar to a rule of thumb that occurred to me about hand-contoured maps: that, whatever the dip and whatever the scale of the map, people tend to choose a contour interval that puts the contours about half an inch apart. This "rule" applies to contouring by hand, but not, of course, to photographic reductions of hand-contoured maps. A similar rule could be made to apply to bands of color, but with a closer interval, because maps on the screen are more like the reductions of hand-made maps than like the originals.

In assigning colors, there are two requirements that you need to think about. One is that the colors should be in some orderly progression, so when you glance at the map, you can easily tell the deep parts from the shallow. The other requirement opposes that one. It is that there should be enough contrast between adjacent colors to make it easy to see the difference.

The first requirement would seem to be one that could be taken care of by putting the colors in the sequence of the spectrum. But that isn't the arrangement that people seem to choose.

The need for contrast calls for something other than a straight sequence of colors with each one next to another that is only slightly different. On

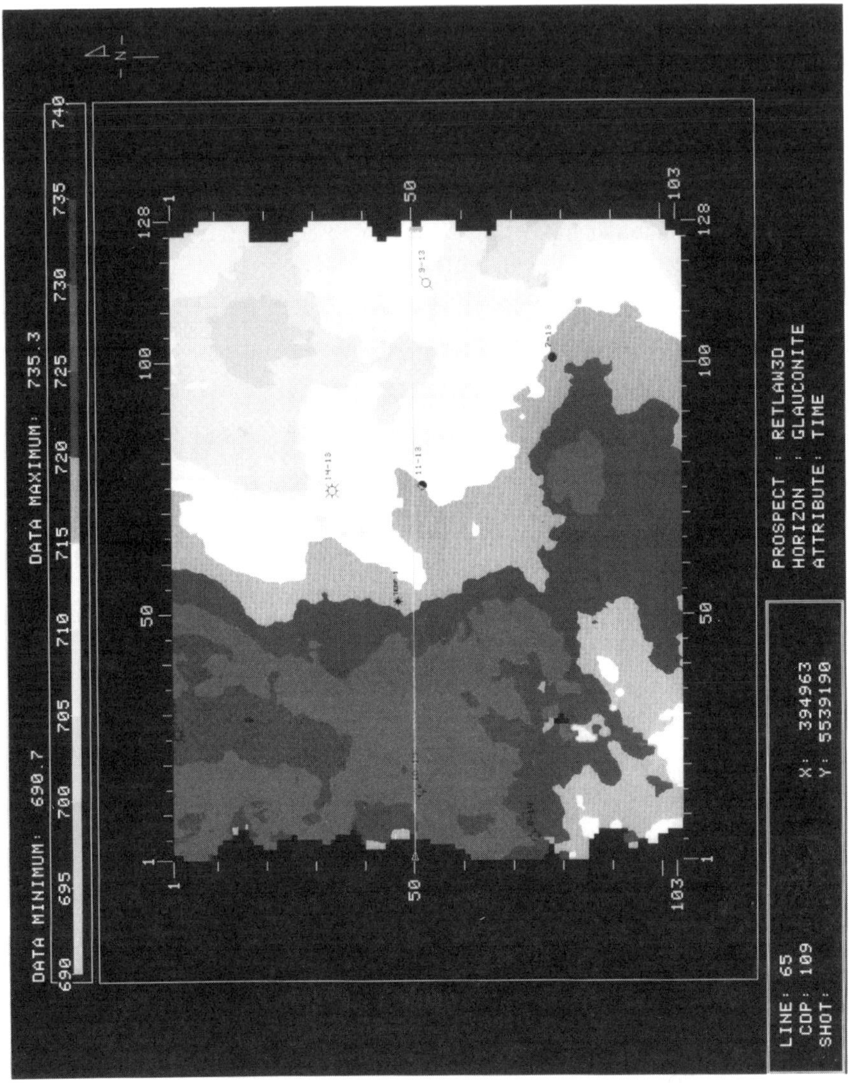

Sierra Geophysics, Inc.

a map it is useful to be able to see contrast between the zones of color. But if your color arrangement is just the spectrum from red to violet, then the changes are gradual. For instance, there may be so many steps that several gradations of green are used. It wouldn't be easy to distinguish between them. So people tend to devise a system of colors that in some places puts contrasting colors next to each other. A scale of the colors and the times they represent is always displayed as a bar on the screen, so whenever you need to, you can check to see what range of times a certain color represents. Mostly they tend to use what I call a "geographical" sequence. It is easy to remember because of its use in atlases, a system that reminds people of the colors of the earth at various elevations. You probably already remember the sequence from maps of the world, continents, or countries: dark blue for the depths of the sea, lighter blue for shallow water, green for grassy lowlands, brown for higher elevations, white for snowcapped mountains. It has two colors that are not in the spectrum, white and brown, and it omits red, orange, and some other colors of the spectrum. You can find places for those colors where they can break up the geographical sequence with some contrast, but still leave it fairly familiar. Naturally, any color scheme you like that helps make reflection times clear to you is satisfactory. The sequence of colors will quickly become familiar to you, so you will be able to distinguish a high point on the map from a low one. And if you want to see a particular contour line, you can assign black, or some other outstanding color, to a very narrow range of times. Then that color looks like a contour line on the map.

What do these color representations do to make a map understandable? It depends on how much they fill in the area. The bands of color are usually about a millimeter or so wide, thick enough to be easy to see. If you are working a 2-D survey, they make up a grid of vari-colored seismic lines. The colors aren't the same as contours, as they don't fill in between the lines. That's what contouring of scattered control is—guessing between the lines. But you can readily see the regional dip, by noticing what colors are dominant near different edges of the map. In a 3-D interpretation, if you are just working a loose grid selected from the lines available, the usefulness of the color is about the same as it is in 2-D.

But when you completely interpret a 3-D area, or if you use a map-interpolation feature to fill in automatically, then the area is filled in solid with color. The lines are wide enough to meet, leaving no gaps between them, so there are no open spaces that would have to be filled in with contours. The map is essentially contoured, with the boundaries between colors acting as contours. So you are looking at a contoured map of your interpretation on the screen, more or less as you pick. It doesn't display faults as such, but just shows the throw of a fault as steep dip. Depending on the system, the map may update automatically, changing every time you make a change, or it may wait for you to ask for an update. Each has advantages. The automatic-change system is no trouble. But update-on-request lets you continue to see a sensible map while you try some picking you may be afraid is foolish. Then when you like your picks, you can update the map.

When seismic times are displayed as colors on a map, on some systems the range of colors is stretched as far as necessary to cover all the times on the map. So if there isn't much variation in times, as there won't be if you are mapping a pretty flat horizon, all the colors you are using will be applied to that short range of times. The map will be colorful with variations that represent small differences in time. That is just the way you want it if you are mapping low-relief features on that level horizon—the gentle features show up in color differences. If you are mapping an area that has steep dip, then the variations in color occur only on features that have high relief. That too is the way you want it. But there is one caution here: the colors change as the range changes. You will notice the effect most in an area with considerable overall dip.

As you start to work it, maybe starting in a restricted area around a well, the whole range of colors will show the dips you have picked. But as you go farther from your starting point, the regional dip will give you points higher or lower than you had already mapped, so the colors are more stretched out, and a feature that had at first stood out in color now has little or no difference from the color of its surroundings. It's not a bad problem, but you can be fooled if you aren't thinking about it. If you want to avoid the problem, you can start your picking with two picks, at the highest and lowest points of the horizon. Of course, the reflection for that horizon has to be very good for you to determine at the start what its highest and lowest points are. With those two points at the start, the system will spread its colors between them, and all the other points you pick will be between them, and so will be assigned colors that will not change as you pick.

POSTING AND OVERPOSTING

Instead of just representing the times by colors, you may be able to have the times posted as numbers on the map. For that to be useful on the screen, you would need to be looking at a small area, maybe a small portion of a larger area. When the time comes to make a large map, perhaps for a report, the computer can send the seismic times or other quantities to a plotter, and post them on a film or paper map.

Computers often overpost, plotting one number (or word) on top of another, so both numbers are hard to read. Usually when a computer plots data on a seismic line, it does not consider other numbers that may have been plotted on other lines. The shot point numbers and reflection times are plotted in uniform relation to the shot points. For instance, each shot point number may be placed above and to the right of its shot point, with the digits extending perpendicular to the line of shot points. Then the numbers are similarly plotted for another line. If one of the numbers on the second line happens to fall on one from the first line, or on a shot point from the other line, it just does. What a person drafting by hand does is to put an individual number in a different relation to its shot point, around on the other side of it, or somewhere, so it doesn't run over any other number or shot point.

When one number falls right on another, maybe neither can be read. A person drafting that way consistently would be fired, so why do we let a computer do it? We do because the computer has other advantages that outweigh that disadvantage. A computer that worked slowly and made mistakes like a person does would be shut down for repair or traded in for a new one. And since the numbers usually aren't used for hand contouring, some of the need for having them legible is gone.

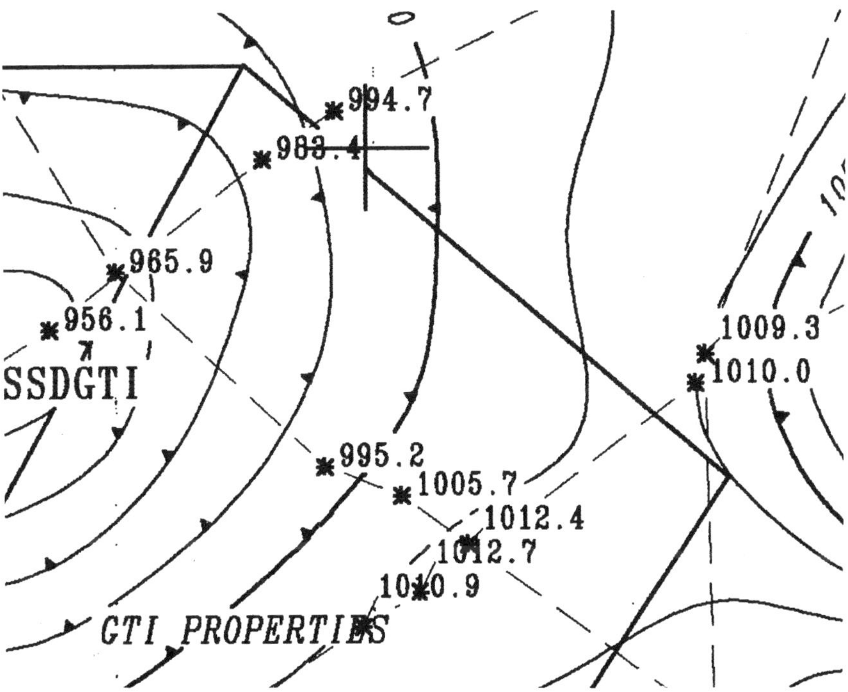

Geophysical Techniques

Fig. 4–6 Overposting avoidance

There are two ways to make the numbers not overlap. One involves human interference. Using a standard drafting program, a person can move reflection times and shot point numbers one by one until none of them overlap. In the other method, a mapping program may have the ability to avoid overposting. The system can't see what's been plotted like a person can, so instead it checks its memory to determine whether something has already been put there. If there is something there, the new number is moved to a clear space nearby. To keep the number associated with the proper point, it can't be allowed to move the number far though, so if the vicinity is crowded it may have to let two numbers overlap to some extent. It results in a more legible map than one with uncontrolled overposting (Fig. 4–6).

COMPUTER CONTOURING

Contouring is traditionally done by hand. It allows you to think about data as you contour it; it makes you aware of every bit of the contouring; it lets you make decisions when gaps in the data leave room for alternate ways of contouring. And it is fun. Getting to contour can, in 2-D interpretation, be a reward you give yourself for picking sections, making all the loops tie, and adjusting lines to one another. But when there is a lot of data to contour (many horizons or a large 3-D area) the fun can degenerate into tedium.

As in so many other aspects of seismic interpretation, large amounts of data get to be too much to handle manually, so computers take over the task of doing something with all the data.

Computer contouring (Fig. 4–7) has some advantages that derive from the ways computers handle data. It doesn't make the kind of mistake that a person makes by misreading a number. If you give the same set of data to the same contouring program twice, with all the choices within the program the same, it will produce exactly the same contours both times.

The way it handles data also gives it some disadvantages. It can make some real blooper errors in large gaps in the data or at the edges of the area. To get it to make alternate contourings of an area, you have to do something to push it into a different form: give it different options, add some artificial data points in gaps, or re-draw parts of the contouring by hand.

The color display that shows on the screen is only a kind of pseudo-contouring. When a 3-D area is interpreted without skipping lines, then the color display is like a contour map, with the changes in color where contours would be. But that simple method doesn't handle faults well. And in 2-D, or in 3-D with some lines interpreted and others skipped, it is not at all a contour map. It just has different colors that indicate ranges

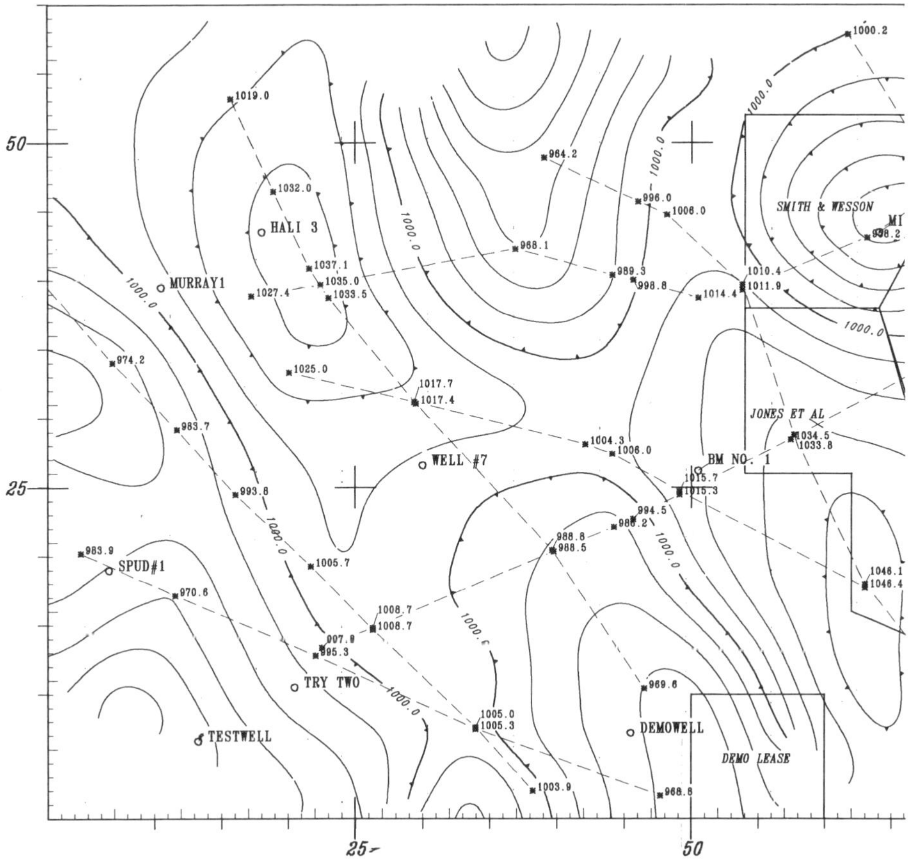

Geophysical Techniques

Fig. 4–7 Computer contouring

of reflection times. Instead of being contours, they are more like plotted times, containing information that you could use to contour the map by hand.

A true contour map is considerably more complex. To produce it, a contouring program is used to draw contour lines. Writing a program for contouring is difficult, much more difficult than learning to contour by hand. A person can contour scattered data fairly easily, but that would be much harder for a computer program. So most of the programs go at it a different way. First they "grid" the data, that is, estimate reflection times at the intersections of a regular grid. The estimates are made by evenly distributing the differences in reflection times between actual data points. So each grid point has a time that implies a smooth gradient between nearby data points. Once there are times on a grid, a program can contour those uniformly spaced grid points.

This thorough contouring may be included in a system or it may be a separate program. If it is a separate program, then to use it you will need a list of commands to give, probably by keyboard. You can memorize the commands, or if you don't use them often, just keep them handy on a piece of paper.

There may be a less-complete (shorter) contouring program on the system. The shorter program may be the only contouring program on the system. Or it may be in addition to a more complete one, just to produce a quick contour map. The short program will probably contour in straight-line segments. Then there may be another command to smooth the contours. This smoothing is just rounding the straight lines into smooth curves, without considering the data.

When there are open spaces between picked lines (in 2-D, or when you skip a number of 3-D lines between the lines you pick) any contouring has to make guesses to fill in the open spaces. That is the main skill you use in contouring by hand. You have to make good judgments to avoid wild-looking contours that swing back and forth between lines, or timid-looking ones that stay close to the seismic line, so the contours follow the pattern of shooting.

A computer contouring program does its gridding by assuming uniform dip between data points. You need to be able to change the contours it makes. If nearby data makes it reasonable to think that a syncline pokes in between two lines, then you need to get it drawn in there. There are different ways to make a change like that. One is to put an artificial reflection time somewhere between the lines, and then let the program contour again. It will include that point in its gridding, so its contours will show the syncline. Another way is for you to be able to change the contours by hand. Give a command that lets you draw on the contour map, and delete and re-draw whatever contours need it. Then you can have a smoothing function change the part you drew in to smooth curves like the other contours. Another way a system may have to put your ideas in the contours is to draw a few contours by hand before using the computer contouring. The lines you draw are taken by the system to be data, and are gridded along with the other data. Then the computer's contouring, in taking those points into account, is like your contouring.

Faults present a special contouring problem. The upthrown and downthrown sides of a fault at a horizon are close together on a map. So a simple contouring program would be likely to make some strange contours in trying to contour those points. Programs for complete contouring have a provision for that. You mark the places where reflections end at faults. That forms a line of points along each side of every fault. Then, in a separate step, you draw a line connecting the points on the upthrown side, and draw another for the points on the downthrown side. Bring the two lines together at the ends. For each fault, that process forms an outline around the fault. The outline consists of two lines, nearly parallel, that come together at the ends of the fault. In drawing the outline, you may not have to connect at the second end you come to, because the program may make that last connection for you. When the map is contoured, the contours stop at the fault outlines.

There are different ways contouring programs handle a fault outline. The simplest is to just contour the numbers as though there isn't a fault, and then blank out the contours within the fault outline. Using that kind of program, you may need to make the fault outline wider than it really is, to cut out the strange-looking contours near the fault. Another type of program contours to the fault outline, stops, and then does the same on the other side of the fault. The most sophisticated contouring does what a person does—contours each side separately but similarly. For instance, an anticline broken by a fault is contoured so you could cut a map along the fault, shift one side, and get the contours to fit.

Overlaying true contouring on the pseudo-contouring of color ribbons makes an effective display. The contours take proper account of faults and other features, and the colors make it easy to see the directions of dip (Fig. 4–8).

WIRE NET

On a map that has contours, color-coding, or both, the highs and lows are easy for you to recognize, but that is because you are accustomed to looking at contours. Their meaning isn't instantly clear to people who don't work with them. It is desirable to be able to show the contours in some way that is more readily understandable. A clay model of the features would of course be easy for anyone to comprehend. Next to that would be some kind of perspective drawing. With the data in memory, the computer can make such a drawing. To make it so all the features are evident, it needs to somehow show more detail than, say, a painting of a mountain scene shows. The technique used is to show a grid of lines in two directions, like a picture of a net draped over the area. It is called a "wire-net" display.

To make a wire-net display to your liking, you can choose the vertical exaggeration and other parameters. Since a perspective view doesn't show the whole area (the far side is hidden by nearer features) you can choose the direction to view from, and the elevation of view. Then, to see the other side, you can display it from a different viewpoint. You are looking

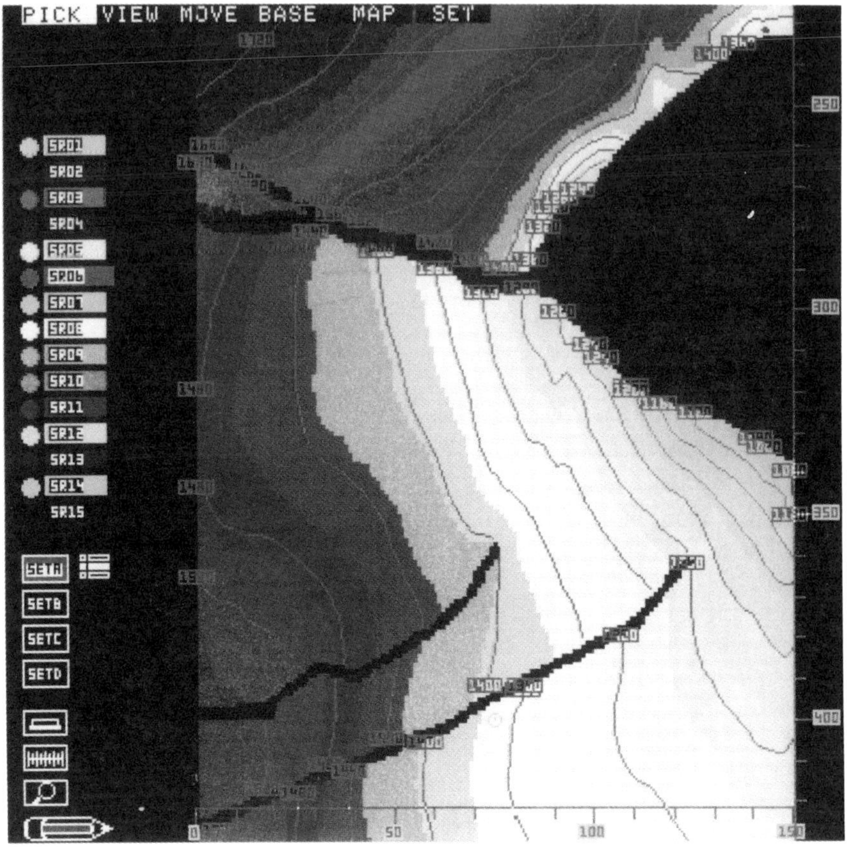

Graphic Science Corp.

Fig. 4–8 Color-coding with contours overlaid

at it the way you would look at a clay model, turning it this way and that, looking from above or from the side, but the variation in vertical exaggeration gives you different views you couldn't have with the solid model.

To combine the advantages of contours and wire net, a contour map can be displayed as though it is on a flat layer above a wire-net display. It too is in perspective, so you see it as though from an angle (Fig. 4–9).

SHOULD CONTOURS HONOR DATA?

The answer to the question of whether contours should honor the data seems obvious. At first a person would think that a simple "Of course" would be the only reasonable answer. But seismic times are notoriously

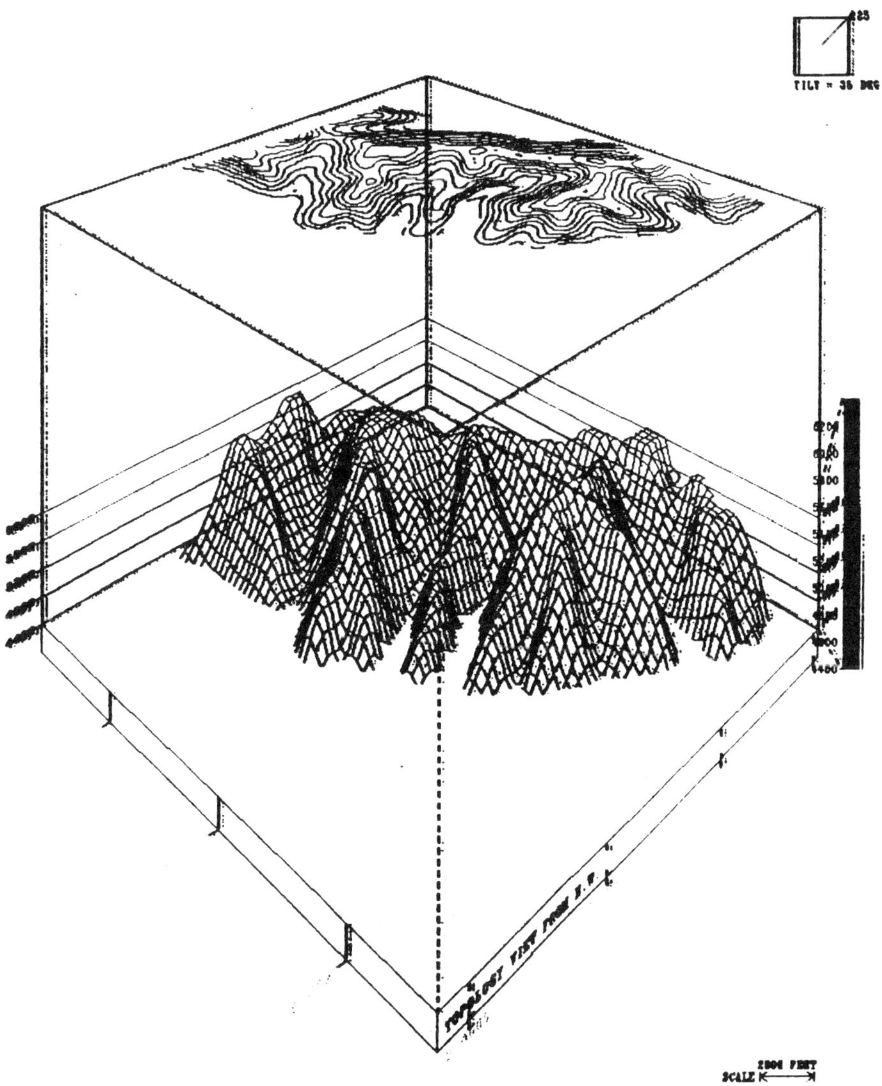

Geophysical Techniques

Fig. 4–9 Wire net and contours

vague. They can be affected by near-surface problems that aren't completely corrected in processing, by velocity variations, and probably by a number of other factors. So contouring the data exactly would lead to strange-looking contours representing shapes that the rock layers couldn't possibly have. Some smoothing is necessary in contouring seismic data.

In contrast to that, hand-contouring of well data is usually slavishly precise, not only honoring the tops at the wells, but carefully putting a contour at the proportional distance between wells that it would have if there was straight, uniform dip between wells.

In contouring seismic data by hand, people have various ways of smoothing. Sometimes there is a risk of the smoothing's going too far, with preconceptions influencing the contours more than the data does.

Now comes computer contouring of seismic data. Should it honor the data exactly? Well, no, unless you want it to have the funny squiggles that a person avoids in contouring by hand.

Most computer contouring uses a method that doesn't strictly honor the data anyway, thus fitting neatly with the need for some smoothing. The contouring is done by making a grid of points to be contoured. The way it is handled is as though you laid a square grid of points on a map of seismic data, then, at each grid point, put a reflection time that you determined by some sort of averaging of real data points around it, then threw away the real points and contoured the grid. It isn't done that way, of course, but by calculations in the computer that give the same effect. There are variations of the method depending on how much more weight is given to near data points than to farther ones, and what is done at grid points that don't have nearby seismic data.

One form of contouring, triangulation, doesn't grid, but instead calculates triangles from data point to data point. Triangulation honors the data exactly, but doesn't do very good contouring. It is fast, though, and is used by some systems to make simple maps quickly.

CALCULATIONS

When you pick horizons interactively, the reflection times at all the traces are automatically determined and stored in memory. With all those numbers already on hand, and with a computer's ability to do arithmetic speedily and accurately, there are some useful products to be had, just by getting the computer to make some calculations. So some programs have commands to perform computations that are useful in seismic investigations. A system may have a set of calculations it can perform on the picks for entire horizons, that is, making a calculation for each point on the map.

A simple calculation is just adding a constant to each point on a horizon. This is a quick-and-easy way to make the picks tie wells. Maybe the reflection from an important formation is poor and hard to pick reliably, so you pick a nearby reflection, say 37 ms shallower than the one from the formation, a better reflection that gives you data you can trust. Adding 37 ms to all the times will produce the map you would have had if the reflection you wanted had been better.

Another bit of simple arithmetic is multiplying the times of a horizon by a constant. This can be used to produce a depth map, in the simple case of a constant velocity to the horizon. The constant to be used is not the velocity though, but half of the velocity, to make up for the fact that the reflection time you are using is two-way time, down to the reflection and back up.

Subtracting the times of one horizon from those of a deeper one gives a time interval map, often called by the misleading term "isochron" or "isotime" (they mean "same time", and really apply to any time contours). A time interval map is like an isopach map, but in time. Such a map is useful in several ways. Where near-surface velocities are difficult to correct for, it allows you to map a reliable shallow horizon (shallow, but below the problems), and subtract it from a deeper horizon to determine an approximation of the structure of the deeper horizon. What you are doing is assuming that the shallow horizon is flat. It is likely to be nearly flat, so that error isn't great. And you are avoiding the distortions and uncertainties that the near-surface problems would cause in the deep horizon. The difference between two horizons also approximates the structure of the deep horizon as of the time the shallower one was deposited, and so is useful in working out the geological history of the area.

You can use the ability to multiply one horizon by another to obtain a map in depth. For the purpose, any set of numbers at points over the area can be called a horizon. Start with two horizons, a regular horizon in reflection time and one consisting of velocities down to the reflection. Multiply one of the horizons by the other. Multiply the result by a constant, in this case .5, because you want to get the distance down to the reflection, not the down and up of the two-way reflection times.

Or the time horizon can be a time interval between two horizons, and the velocity horizon can be the interval velocity between those horizons. That gives the thickness. It can be used to do a peel-off calculation, determining the thicknesses of the various layers and then adding them together to get depths of the horizons.

Dividing one horizon by another doesn't sound so useful, but there are things you can do with it. For instance, you can make up a horizon by interpolating between well data for a formation, then dividing that by the times to the reflection from that formation, to get a velocity function to that formation.

A calculation can be made to smooth the picks made on a horizon, so it makes a smoother map. This brings you nearer to the truth in a seismic project in which some problem in shooting or uncertainty in picking, causes your horizon picks to be irregular.

This performing calculations on the data is one of the powerful capabilities the system gives you, along with others like the automation of picking, smoothing, and contouring. But your use of all these abilities can be limited by your access to the system.

YOUR TURN ON THE SYSTEM

Like many other activities, working on an interactive system can often be either feast or famine. However many systems your company has, one week they are sitting idle and management is wondering why you interpreters talked them into getting so many (even if "so many" refers to just one); the next week you are taking turns and waiting to use them, and people find it strange that you aren't better occupied. The different situations are often predictable enough. When a lease sale or an annual budget meeting is imminent, all of the interpreters are in a rush to come to conclusions about prospects, and to get reports finished. And when the crisis is over is a good time for you to get information together to start on new projects. You aren't ready to use a system heavily.

Just when you are ready to use it, someone else may have started interpreting on it. Or there may be a schedule of times when it is available to you, and your needs for it may not fit that routine. One way or another, you are likely to have to fit your use of it into other people's plans. Usually, letting the interpreters plan a schedule of sharing a system among themselves seems to work better than having a regular schedule of times. They can check with one another and decide who needs it most and when, more smoothly than any rigid arrangement. Even so, after you do sign up for it, there can be little misunderstandings that bring your schedule into conflict with someone else's.

If the decision of when to have your turn at the system is yours to make, you may wonder what time to choose.

NIGHTS CAN BE BETTER THAN DAYS

If your system shares a computer with other terminals, it can sometimes be pretty slow in its part of the interaction with you. If you work a little past quitting time, you may notice that its reactions are quickened. What a shame—just when the work gets easier you are shutting down for the day, to come back next morning to competition again. The answer to that is simple: work at night. Companies normally want their employees working at the same time, so they can receive instructions, confer with one another, and make reports to the boss. But if there is so much competition for computer power that the work is slowed considerably, there is also likely to be competition for a place at a workstation. In that situation, the company may ask interpreters to work nights. The trick is, if you want quick response from the computer, to be one of the interpreters who doesn't object to working at night. That brings the other advantages of not being interrupted and of having the place pretty much to yourself. Sometimes you'll be amazed at how much more work you can do at night than in the same number of hours in daytime.

BUT NIGHTS CAN BE WORSE THAN DAYS

There are some situations that can make working at night less effective than working in daytime. People plan their data-loading and other batch jobs to mostly take place at night so those activities won't interfere with work during the day. So you might find that you have more, instead of less, slowing of your work at night. One side effect of night work is that people may ask you to perform some minor tasks on their batch project. You may be asked to take some simple action when an error message appears, give a few commands to start the next batch going, etc. There can be a slight advantage in that it familiarizes you a bit with the loading or other batch process. But in either that or your own work, there isn't anyone available to help you if you run into a problem. You may feel responsible if the computer breaks down, although just operating it isn't likely to be the reason for the breakdown. And the cleaning of the offices goes on at night, just so it won't bother people working in office hours. Waste baskets are emptied and vacuum cleaners roar for a while. If you are a consultant working on the system at night either because it is cheaper then or just because it is available then, you won't be able to show displays on the screen to your client. You'll have to do it all with hardcopy.

WEEKENDS CAN BE WORSE THAN WEEKNIGHTS

Another time you might work on a system outside regular office hours is on weekends. For other kinds of work, weekend time, often Saturday morning, is well known as a time to get good work done. But weekend work on an interactive system runs into a problem about air conditioning, as I did (and described in the segment "Computer Problems"). Managers of office buildings naturally don't want to use a lot of expensive power running air-conditioning equipment at a time the tenants don't normally work. The lease may even state that heating and cooling will not be provided at night or on weekends or holidays unless specially requested and paid for by the tenant. And your company, or its accountants, may not be keen on paying to keep a whole block of offices cool for the comfort of an employee or so, who may or may not accomplish anything. And on top of all that, even if your company requests cooling and agrees to pay for it, the building's maintenance people may not get the word. If air conditioning is turned off at night, nights are cool anyway, so the building may not get much warmer. But weekends are different.

Normally, the cooling is cut off at the end of the work day Friday. Friday night the building doesn't warm up much. Saturday morning the sun comes up and starts warming the building. But the building is massive. The walls, floors, and furniture are still cool, so they keep the inside air reasonably cool. Saturday afternoon, the heat builds up some more. Saturday night isn't so bad, it's a time of equalizing temperatures. But then comes Sunday. The sun comes up and starts heating a building that has about used up its stored coolness. As the day progresses, the building gets hotter, and the computer's fans are just blowing out hot air to let not-much-cooler air into the machine.

If you do manage to have air conditioning in your part of the building on the weekend, either by a previous arrangement or by your company having its own air conditioning in the room with the systems, the air conditioning may be less effective than normal. In both situations the building air ducts are used, and the blowers for just your part have to try to push the air all through the ducts without help from other blowers in the building. And other rooms nearby are getting warmer. So your cooling has an uphill fight.

Moral: If you are planning to work on an interactive system on the weekend, do it Saturday, preferably Saturday morning. Computers can't stand Sundays.

Why do mainframe computers, bigger, producing more heat, and more sensitive to heat, run 24 hours a day, seven days a week? They do **because** the problem is severe, so proper allowance for it is made. They have continuous air conditioning, appropriate ducting, and backup generators of electricity to keep the air conditioners running if the power fails.

SMALL SYSTEMS AND OTHER ALTERNATIVES

There is another way to handle the access problem: by having your own system sitting ready on your desk. Some manufacturers of interactive systems have set themselves a goal of providing a workstation for each interpreter. But the complete systems haven't yet reached that level.

The goal of having an interactive system for each interpreter won't be achieved until the systems go considerably further along the path that pocket calculators have taken. In 1972 I bought a calculator in Singapore for the equivalent of about US$125. It was bulky enough to just barely be in the "pocket" category. A few years later pocket calculators were selling for around $30. Now they are still cheaper, and some have the length and width of a business card, are thin enough to slip into a shirt pocket along with a note pad and pens, and run for months on a battery or even on solar power.

Like calculators, interactive systems are getting smaller and cheaper, especially as the computers they are based on become smaller. Even if your company does have the latest in interactive systems, there may not be one available to you when you need it. There are some alternate ways of continuing with your interpretation.

SMALL SYSTEMS

Some systems are based on personal computers (Fig. 4–10). That makes the systems considerably smaller and less costly than the standard systems. The small systems perform some of the functions of larger ones, but with less memory and speed. For some small systems the personal computers may be boosted with an additional computer unit added or a special interface card fitted in the computer.

The small systems have their own advantages. The company can have enough for them to be readily available, maybe one for each interpreter, and so can get by with fewer large systems. You can perform many of the tasks of interpreting at any time. A small system can be cheap enough to be allowed to stand idle when not needed, like a typewriter or a copy machine. And with its smaller size, it fits in a normal office, on one end of a desk.

Seismic Micro-Technology, Inc.

Fig. 4–10 Small system

A small system may be able to calculate various seismic attributes and may display sections with the attributes shown in color. It may be able to flatten horizons and let you pick on the flattened section. You can derive wavelets from the seismic data, and use those wavelets to make synthetic seismograms for identification or modeling. The system may let you try some processing procedures like muting field records, and in other ways help you to investigate the nature of the data.

Operating a small system is about the same as using a large one, except that on a small system you may give more of the commands by keyboard, pressing a key to select a command from a menu (Fig. 4–11).

A system may store data on floppy disks, one or two seismic lines to a disk, or you may be able to put some lines on its hard disk. It would be difficult to pick a whole area on such a system, but you can use it for automatic or semi-automatic picking of selected parts of lines when analyzing problems. These analyses and different kinds of displays work as well for 3-D areas as for 2-D.

There are many advantages to being able to do at least some of your interpreting on a computer that is on your desk ready to use. You not

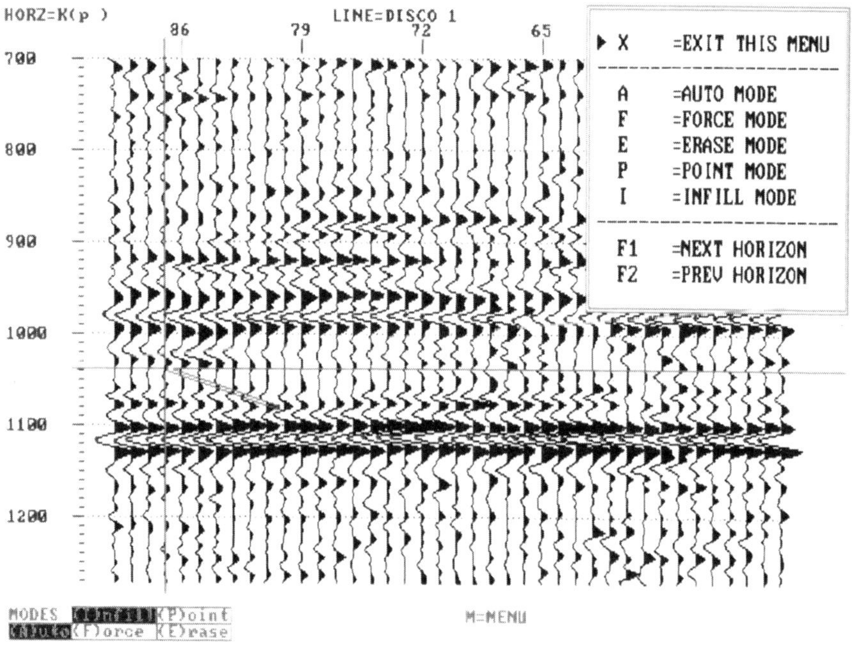

GMA

Fig. 4–11 Menu for selecting by keyboard

only don't have to wait your turn, you also don't have to leave your office. You can still answer your phone, look up notes, etc. Any time you want, you can call up a section, look at it in different attribute displays, and leave it on the screen as long as you want. This fits the way the mind works—the most effective time to test an idea is as soon as it occurs to you. Having the computer handy makes it easy for you to play with the data, to try different displays (Fig. 4–12) and different picks just to see what they might show. That kind of experimenting can have great benefits. And having a personal computer handy is convenient for other reasons. It isn't limited to interactive interpretation, you can use it for other things like writing reports with a word-processing program and performing calculations with a spreadsheet.

Even some portable computers are powerful enough to run a small system. You can take a portable along on a trip, with several seismic lines on floppy disks or on its hard disk. The displays of various attributes will probably not be in color. Color displays are just beginning to be available on portable computers. But without color, you can display as though in color, letting shades of grey take the place of the different colors. You can do some picking, and have the picks ready for transfer to other equipment.

The normal systems are becoming cheaper, and the small ones more powerful, so the two are approaching each other. And as in the development of any invention, at the same time that smaller, more powerful ones are being made, much more powerful large ones are also being developed.

Surprisingly, one of the main advantages of a small system can appear when you are in a rush, and need to get work done sooner than you could on a larger system. Back when all computers were mainframes, and most of the work done on them was in batch form, people made much of how fast a computer could operate. A task that might take half a day to do by hand could be done in seconds by a computer. But when we had something done on a computer, we often got the finished product a week later. If we were in a hurry, we did the work by hand. The distinction was that computers were faster, but manual calculations were quicker. Now the same principle applies between large and small interactive systems. For some functions, a small system may not work as fast as a large one, but if you have to wait for a turn at a large one, or even wait to get back from an out-of-town trip to use it, you can get the work done sooner on a small system where you are. **There is no substitute for a powerful computer and large color screens, but there is also no substitute for immediate availability.**

PAPER SECTIONS AND HARDCOPY

Another way of continuing your interpretation when the large interactive system is not available for you to work on is with paper.

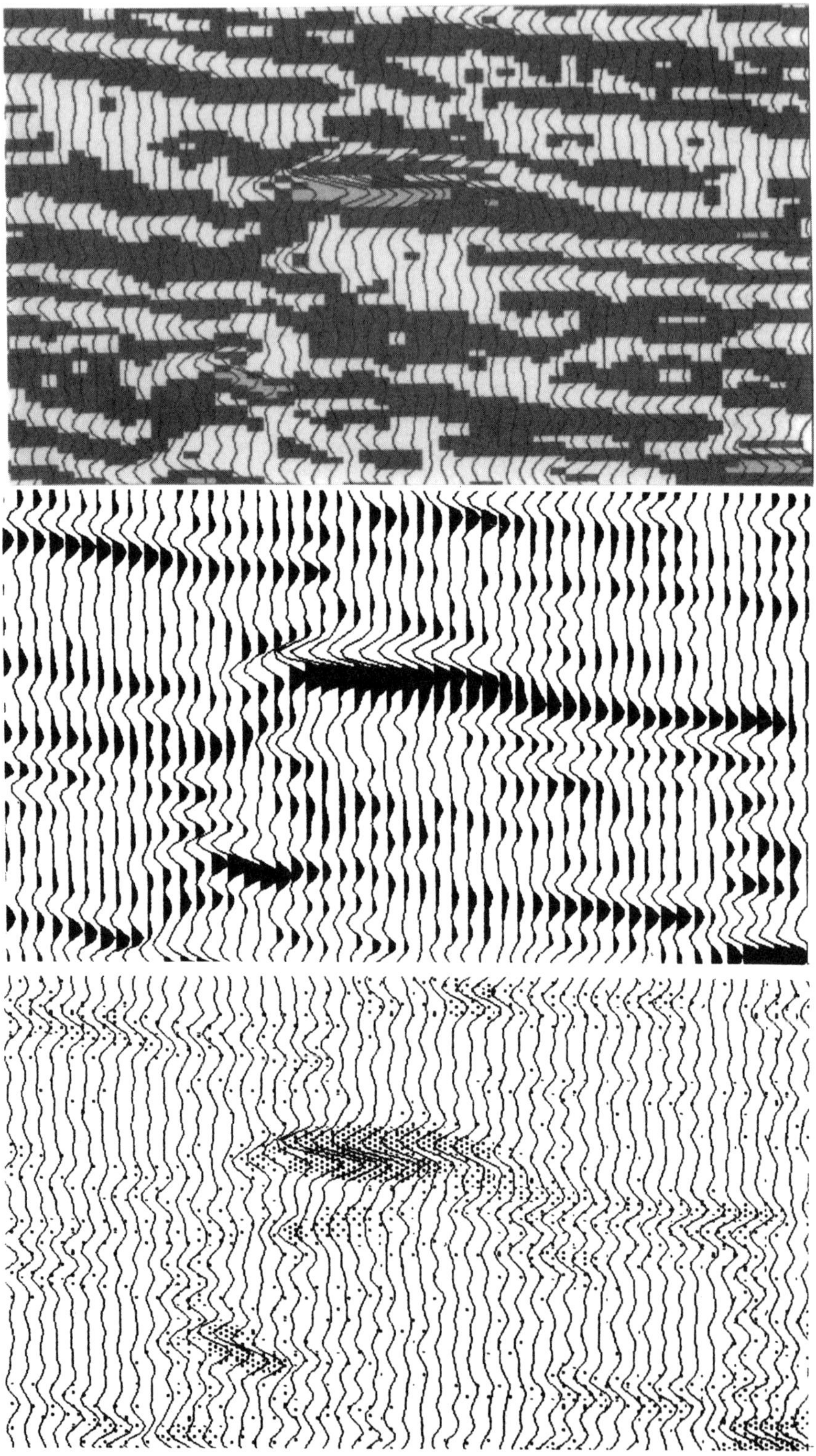

Seismic Micro-Technology, Inc.

Fig. 4–12 Bright spot

The use of paper sections when using a computer to interpret seismic data was discussed in Chapter 1 in "Transition Period". But instead of working with old-fashioned paper sections along with the interactive system, you may sometimes find it convenient to work another way around, to interpret, back in your office, on paper hardcopy you made on the system. Maybe you are trying to make up your mind just how some deposition occurred. With the system, you can flatten on various horizons that represent times immediately following stages of deposition. You can display those flattened sections in different attributes that accent different features: amplitude to highlight the bright and dim spots, phase to show continuity unaffected by amplitude, frequency to show compaction of sediments, etc.

Each of the displays you make can be studied on the screen, of course, but there may be limitations to their usefulness there. The system may be in such demand that you don't want to spend much of your limited time on it, sitting and mulling over a section. The size of the screen limits how many sections you can compare easily. Of course, if you have two screens, you can show one section on one screen and another on the other. And you can put several sections in windows, but those have to be small pieces of sections to fit in. So there isn't any way to get quite the convenience of shuffling a few printouts on your desk, and marking tentative picks on them. You can make the copies during your time on the system; then go into your own office; and, at your own pace, look through the sections you've made and compare them. You may come to conclusions that will help you when you go back onto the system, and let the system do the legwork that is needed in order to follow up those conclusions.

Not only will you be able to use the paper printouts away from the system, but you can use them in conjunction with old-fashioned paper sections, spreading both kinds of section out, comparing them, and making picks on one with guidance from another. To use the printouts in place of screen displays, you may need to have timing lines on the hardcopy, to substitute for the handy locator that tells you where you are on a screen display.

Sometimes you may not even want the display preserved in the computer, but just shown on the screen so you can make a paper copy of it. That seems improvident. What if you should need it again? The answer is that it isn't forever unavailable, you can just have the system recalculate it, so it's a matter of balancing the time wasted re-doing it if you need it again against the disk space wasted storing it if you never need it again. The decision partly depends on your system. If you have a powerful computer, then you can quickly remake a special section, so there's no use complicating your list of items in memory. If your computer is less powerful and therefore slower, and if you have plenty of memory, then it may be better to save the sections just in case you want them later.

PRESENTATIONS

One important aspect of any seismic interpretation is the presentation of results. The usefulness of the best interpretation is diminished if it is poorly communicated to other people, particularly people who are to make decisions about it. Among other things, "Neatness counts", as school teachers say. Besides neatness, you need completeness and accuracy. You also may need some sales tools to make your results easily understood and even appealing. Fortunately an interactive system takes care of some of those requirements of a good presentation for you, and helps you with others.

The neatness is automatic. The system doesn't know any way to make displays and printouts other than neatly. Sections and slices are lined up properly, labels are in neat rows and columns, there aren't smudges.

Any section you show to bosses or other people should be labeled with area name, line number, shot point numbers, reflection times or depths, and type of display. That keeps you from having to try to remember, getting mixed up, feeling foolish. Also the labeling tells people what the section is when you aren't around. Working on paper, you know the need for labeling, but it's a lot of trouble to do it thoroughly on everything. On an interactive system, all you have to do is be sure those labels are on the sections that are displayed on screen, and then they will automatically be on printouts you make to show people. Even seeing that they are on the screen displays may just be a matter of not changing the defaults. Anyway, if you see to it once, then all your displays will be correctly labeled with no further effort on your part.

The system won't make mistakes plotting data on a map, so you won't have to apologize and go hunting for the correct number. It may, though, plot one number on top of another. Some companies prefer to leave seismic data off the contoured maps anyway, so for them any overposting is a problem only for shot point numbers.

In some companies, managers have computer-made maps re-drawn by hand. That seems strangely backward and old-fashioned, but it is reasonable enough. A farmout, request for dry-hole money, or other deal may depend in part on the other party's perception that the prospect is well thought out. A computer-generated map may seem to have been dashed off quickly, and so can give a poor impression, even though it may actually be more accurate than a hand-drawn map. The poor impression is especially likely if the person it is shown to is not accustomed to working with computers.

FINAL REPORT

When you have completed an interpretation, you will probably need to make a formal report on it, usually called a final report. There are several purposes for the final report. It presents the interpretation to your superiors. In some organizations, you might be able to just tell them about it, but you would still need a thorough written report on your interpretation, to preserve the knowledge you acquired with the time and effort you spent interpreting. You will rapidly forget much of the detail as you work on other areas. So if you later need to re-interpret (maybe because new information was acquired from more shooting or from the drilling of a well) you will be able to refresh your memory by looking up the report. And, of course, it will be necessary if someone else re-interprets after you have moved to another office.

So the final report must contain all the hard-won knowledge that is necessary so someone unfamiliar with the area can start an interpretation of it. The report should contain all the simple statistical information, like when the area was shot and with what techniques, how it was processed, the well logs you had to track down. All that information is already stored elsewhere, in the reports of the field crew and in other documents, but those other sources are scattered, so it would take time to track them down, and later on some of them may become hard to find. Maybe the area will be given to another company to interpret so they can decide whether to join in a drilling deal. They won't have these other sources available.

The report should also have what you have discovered in the process

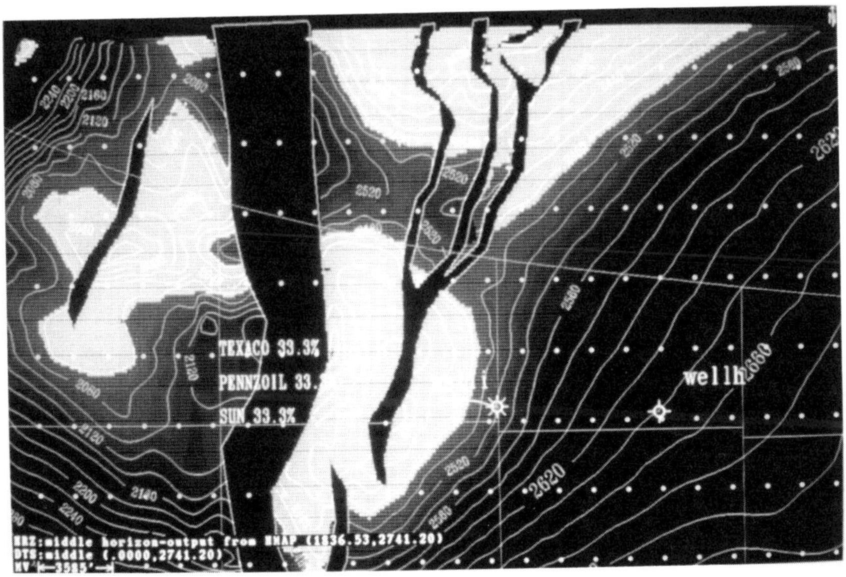

Landmark Graphics Corporation

Fig. 4–13 Map for final report

of interpreting: about the nature of the reflections, special problems and ways you handled them or ways that failed to resolve them, etc. It should have the maps you made (Fig. 4–13). And of course, it must have the conclusions you came to. They were the reason for your doing the interpretation in the first place.

The statistical information is something you have to look up. Some of it you will need in your interpretation, and some you will be gathering just in case someone needs to know it later. But your writeup of the characteristics and problems of the area will come from the experience you gained while doing the interpreting. To illustrate some of the points you make, you will need examples of sections, slices, maybe synthetic seismograms. This is where you will need the picture and annotation functions. It's best to save these examples as you encounter them instead of waiting until you are putting the report together. Saving them as you discover them, you will save too many, but it's easy enough to ignore or delete the ones you don't need.

When you have a particular problem or see some interesting phenomenon, put a good example of it on the screen, and annotate it with simple, plain-language labels. The main use for the labels at this point is for yourself, so you will later know why you saved the picture. At report time you may want to call up the section (or whatever it is) again and make a new picture to print for the report. You probably won't need to make it more formal, but might need to explain more thoroughly, with information that you know but that another person might not be aware of.

If you are using a small system, based on a personal computer, it probably won't be able to save part of the image on the screen in a "picture" file. But you can make a screen dump, printing out whatever is on the screen, the commands included with the section. As you work, you can save these prints in a notebook or file folder. To polish up a screen dump for the final report, just cut away the parts you don't want, with scissors, and tape the illustration on a page. If you want to be fancier than that, make a color xerographic copy of the pasted-up page. Another use of color copying is to provide several copies of your report, for instance to distribute to your company's partners in the area. The color copies can make the report look more polished, more like published material, than pieces cut out and fastened to pages.

This color copying is also a good way of making multiple copies of color illustrations if you are using a larger system. You may not want to tie up the system just to make several prints just alike. It's like the text. A secretary using a typewriter doesn't re-type to get more copies, and one with a word processor may not print out multiple copies. They both can use an office copier.

In some cases, though, it may be cheaper to make the copies one by one on your color plotter; it depends on the plotter. The copies made on some are expensive, and those made on some others are considerably cheaper than color xerography.

For some illustrations in reports, copies of color printouts can be made in black and white on a regular office copier, and still be quite satisfactory. Much of the information in a color display is not added by the color itself, but by the attribute shown. An informative color display, copied in black and white, often shows the significant information almost as well. I discovered that one day when I made a black-and-white xerographic copy of a color hardcopy of a section. You can check this by selecting a color hardcopy that shows some feature you wouldn't have noticed on a standard section, and copying the hardcopy on an ordinary office copier. The discrimination between high and low velocity (or extremes of another attribute) may be lost, but in many cases the distinctive feature is clearly visible. You have surely noticed that that principle is used in this book, with the black-and-white illustrations making their points about as well as color, and certainly being more convenient than a collection of color pictures together in one place in the book.

A final report, if you've done it well, will have all the information necessary for someone to evaluate the area. It will be easy for a non-technical person to understand. And it will have information that will save time when a re-interpretation is made. Writing a report also helps you give an effective report orally. Having looked up the pertinent information prepares you for answering questions. Even if you don't remember all the information, you can look it up quickly in the report.

HOW MUCH TIME DID YOU SAVE?

An interactive system is fast. You can pick roughly and have the system snap to the exact point. Timing and posting don't take any time at all. How much faster does that make it, anyway? You could map the structure on a 2-D area with an interactive interpretation system, and compare the time it took with the time for a similar-sized area done on paper. To be fair, you'd have to include the time it took someone to load the data in the interactive interpreting time. But you'd still be much faster.

That kind of comparison would answer a question from management, to justify the purchase of the system. But is it the best way to measure productiveness in interpretation? No, not when you consider why the interpretation is being done. It's done to find oil. Productiveness is a matter of discoveries of oil, not of line-miles picked in a certain length of time.

Once, when "computer" meant mainframe computer, I attended a course given by a computer manufacturer. The company I worked for had just computerized, and the course was to help us get the best use out of the system. It had been obtained primarily for accounting, but it could also be used for calculations for any department that could put its data in an easily understood form.

The computer company's lecturer made a surprising statement, saying that they had formerly told customers that having a computer would allow them to get their accounting done by a lot fewer people, but they had found out that wasn't what happened. Instead, the same number of people did more. That is, the accounting was more thorough; more information was extracted from the numbers. This enhanced accounting was more useful to the company than the amount that could be done by hand and adding machine.

That is pretty much like the way interactive interpretation works. You will expect to do an interpretation much faster. You of course could do a simple structural interpretation of two or three horizons of a 2-D area much faster than you could do it on paper. But that usually is not what you do. With all the power available, you might interpret structure on many horizons. If there is a place where you aren't sure about some choice in picking, you can do the whole interpretation both ways, and see which one looks more reasonable. You can do much more thorough studying of the data, examining bright, dim, and flat spots. You can generate seismic attribute displays to help you decide whether there is gas in the formation. You can do seismic stratigraphy studies. You can flatten horizons to examine paleo-structure.

After doing a lot of those things, you may not complete your interpretation in less time than you would have done the simpler one by hand. You might even take longer to do the area interactively. But you'd know much more about the prospects of the area, and could do a lot better at recommending drilling.

Actually, the time it takes to interpret an area may be dictated more by the time available than by anything about the seismic data. An interpretation isn't something that takes a certain amount of time, because, any time you think you are through, you can probably think of something else that you should investigate. The additional investigations are useful; they do make you better informed about the prospects for oil. Within a reasonable length of time, it would be irresponsible to stop short just because you had done as well as you would have without the system.

The saving is not so much in interpreting time as in dry holes avoided. The extra benefit is not so much in interpreting more areas in the same length of time as in additional oil fields discovered. Be careful about telling management how much faster you can interpret. Explain what kind of interpreting you can do faster and what kind you can do better to find more oil.

5

If It's a 2-D Area

We've discussed aspects of interactive interpretation that apply to both 2-D and 3-D. It might appear that interpretation is interpretation, and that it doesn't make much difference whether the individual lines were shot some distance apart or close together. It might also seem that a system that could handle 3-D, with all its dense control, would surely include all the capabilities needed for 2-D interpretation.

But 2-D data has some special features of its own, features that 3-D doesn't have, features that call for special capabilities in an interpretation system. The differences between 2-D and 3-D data are brought about by differences in the ways they are shot, which in turn lead to differences in the ways they are processed. It is necessary to use different interpretation programs, or different features of a combination 2-D and 3-D program, to interpret them. A 3-D area is shot as a unit and can be interpreted as a unit. A 2-D area has separate lines of data with wide spaces between the lines; it is likely to consist of an accumulation of lines shot in different projects; and some lines may be very long.

THE SPECIAL NATURE OF 2-D

From this nature of 2-D shooting, the main factors that distinguish the interactive interpretation of 2-D data involve:

Gaps between lines
Line length
Line ties
Loops
Migration

GAPS BETWEEN LINES

A 2-D area is usually shot in several stages. The first is often a reconnaissance program, with lines that are long and widely spaced, to cover a large area and provide a relatively rough idea of the features in it. It may be shot in a uniform grid, with lines up to several kilometers apart (Fig. 5–1). Or, on land, the first lines may be just along whatever roads happen to be in the area.

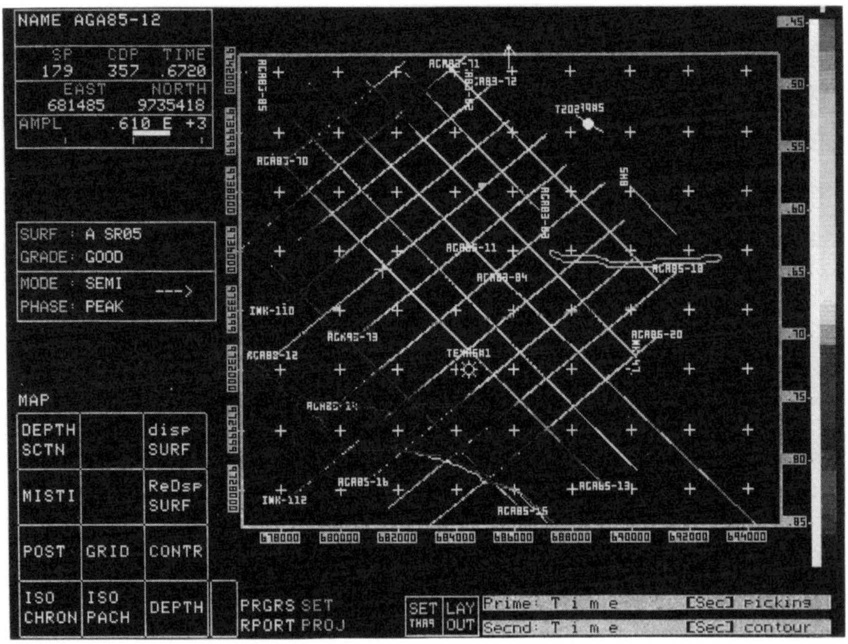

Graphic Science Corp.

Fig. 5–1 Pattern of 2-D lines

After the data from that project has been interpreted, a more detailed program is laid out to investigate some of the features. This closer control may be in lines parallel to the grid lines. But more often, instead of being in a regular pattern, the lines are selected to cross features at critical points and in critical directions. The process is often continued for several stages of closer and closer detail. So the typical pattern is of the initial wide grid with parts of it criss-crossed by lines in many directions.

Most 2-D projects are shot in lines that intersect to form loops, enclosing parts of the area. A few projects are shot with the lines in parallel or radial patterns, so they don't form loops. But whether there are loops or not, a 2-D survey consists of an arrangement of lines with gaps between them. The spaces between lines are generally many times as great as the distances between traces in a line.

Since 2-D data is from an arrangement of individual lines with gaps rather than in a solid mass of data like 3-D, some 3-D techniques either can't be applied to it at all or don't work as well as in 3-D. Lines from a 2-D project can't be called up in as rapid succession as the parallel, uniformly spaced lines in 3-D can. Each 2-D line has to be selected separately. Also, most 2-D data wouldn't be helped by being shown in the form of a time slice, because the wide spacing between 2-D lines would make such a display look more like lace than like a filled-in pattern. However, in a partial way around it, closely spaced 2-D surveys can be used to create a pseudo 3-D area, that can be displayed in time-slice form.

LINE LENGTH

The lines in 2-D seismic projects are sometimes quite long. A reconnaissance survey that covers a large area is made up of long lines. And occasionally a single long line is shot, maybe all the way across a sedimentary basin, for regional information. But there isn't much room on a computer screen. There are several display techniques to partly overcome that problem. A part of the line can be displayed, and the display then scrolled to show other parts.

One way of giving the command to scroll is with a scroll bar, an image across the bottom of the screen that represents the whole length of the line, and has a mark on it to show the location of the part that is displayed at the moment. Putting the cursor at some other point on the bar and clicking changes the display to show that part of the line.

Another way of showing more than one screenful of section is to have adjacent parts of it displayed on two screens. Still another way is to change the relationship of horizontal to vertical, compressing the horizontal dimension until the whole section shows on the screen. Some compressing can be done by choosing a close trace-spacing. Further compression is done by decimation, that is, omitting some of the traces.

LINE TIES

A major problem in 2-D interpretation is tieing lines where they intersect. There are several reasons for picks to fail to tie. First and most obvious, of course, is that you may have picked incorrectly. Another reason is that the lines from one shooting project may not tie those from another. Conventional 2-D shooting has been around for so long that there is a great deal of old data that can be combined profitably with new shooting. The lines may have been shot in different shooting programs. They may have been processed so they don't fit—adjusted to different datum planes, adjusted to the same plane but with different velocities, given different polarities or bandpass filter settings. And sometimes 2-D lines shot all in the same project may mis-tie at intersections. Within the same project, rainy or dry weather between the times the lines were shot may have raised or lowered the water table and thus affected the near-surface corrections. For one reason or another lines may not tie. So you may need to adjust some lines to make them compatible with others.

To tie two lines at their intersection, you can display parts of them together, joined at their intersection point (Fig. 5–2). That is like folding

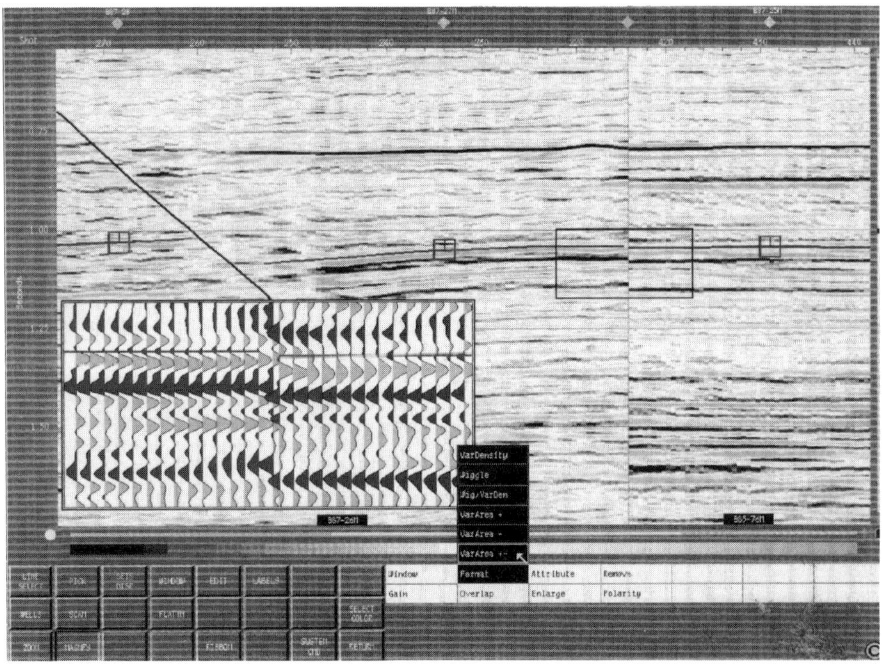

Sierra Geophysics, Inc.

Fig. 5–2 Line intersection

one paper section and laying it over another. On the screen, continue a pick that is already made on one section, taking it across onto the other. The system may not let you make one segment of pick crossing the join; you may have to stop there and start a new rubber band or other form of picking on the other side. If the picks don't fit exactly, shift one section up and down to find the best fit, like sliding the folded paper section over the other one.

On the line you have on the screen, you can have marks appear at points where reflections on already picked intersecting lines meet the displayed line. If the reflection quality is good enough, you can use those marks to show you where to pick the section on the screen, without having to display two sections joined. Or if you have already picked the line on the screen, the marks show instantly whether you have mis-tied, and by how much.

The systems have several ways of showing you on a map how much you have mis-tied. The color ribbons that show reflection time on the map will, by contrasting at an intersection, indicate that there is a large miss. But that isn't very precise. A time on one line near the bottom of a zone of some color and a time on the other line near the top of that zone would appear to tie, although the zone might be fairly wide. Instead of just color, a system may put a number at the intersection, the number of milliseconds of mis-tie, with a plus or minus sign to indicate the direction of the miss. Or there may be a symbol displayed at the intersection to show whether the miss is large or small (Fig. 5–3).

There are several ways to correct mis-ties at intersections. If the error is in the picking, then re-picking is the way to fix the problem. If one line misses all the other lines it crosses, all in the same direction and by about the same amount, then a "bulk shift", shifting the whole section up or down a uniform amount, will make it fit. It would be better to determine why the change was needed, but you often don't have the time or the means to find out just why a line differs from the lines it crosses. If you have lines from different seismic surveys, then one whole set may have to be adjusted to fit another set.

In addition to letting you bulk shift sections, a system may provide other, more dangerous, ways of making lines tie. It may have a provision for averaging the seismic times at intersections. That certainly makes picks tie. It forces them all to tie perfectly at every intersection, no matter how badly they had mis-tied. It does have a valid use. If you have tied closely enough, then it can make sense to average the data at the tie points so a contouring program can handle them smoothly. Computer contouring would have a problem with two times at the same point, but with no geological reason like an overthrust fault. Be careful though, you could conceal a mis-tie from yourself by averaging before you had tied carefully. Another

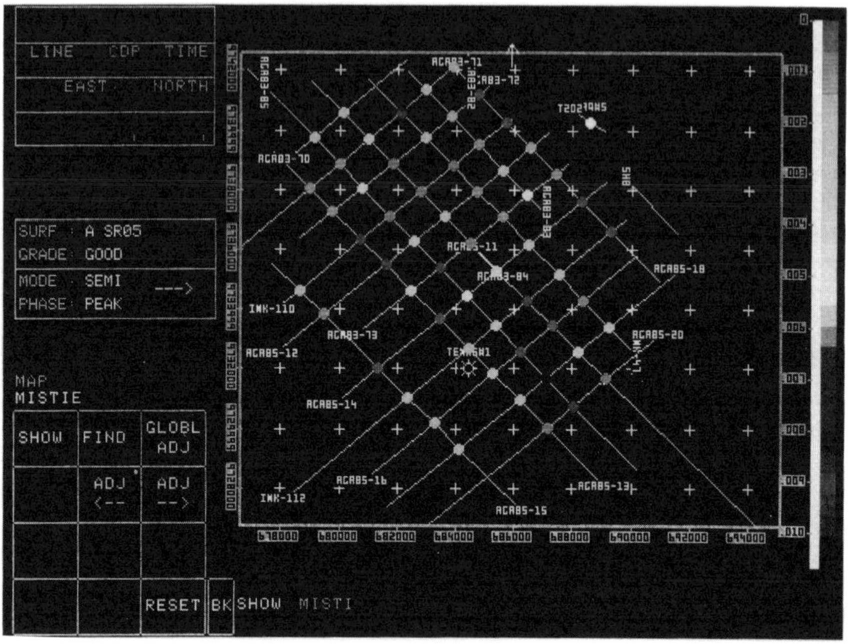

Graphic Science Corp.

Fig. 5–3 Mis-tie map

risky function lets you apply a tilt to a line. That can be useful when you have determined that the line really does require more shift at one end than at the other in order to tie other lines. It would be easy to decide to always apply whatever tilt made lines appear to tie. That would be like overusing the averaging function, but worse. Again, be careful.

To decide how much shift may be needed, you can pick some good, easily recognized reflection over the area, without tieing at intersections. Then look at a map display of mis-ties. The display will show you which lines need what shifts to make everything fit. Or it will if all goes well—some of the mis-ties may be contradictory. Maybe a line crosses four other lines that all tie one another, and the reflection on it is about 22 ms lower than the same reflection on three of the other lines, but is 5 ms higher than that reflection on the fourth line. That's a tough problem that requires investigating the sections, your picks, etc.

In another way of using a strong reflection, you can use flattening to help you decide how much shift was needed, on which lines. Pick the strong reflection without tieing. Then, if your system will flatten a whole area, flatten the picks, and on the flattened sections, pick other horizons.

Time intervals between the flat horizon and the others will show you how the other horizons tie, and will show their structure, except for the omission of the dip of the flattened horizon. The intervals will enable you to determine structure whether you get the tie problems solved or not.

Attribute displays can also help you make lines tie. They show different characteristics of the data that should match at intersections. Reflection strength, for instance, shows blobs of high amplitudes that are independent of polarity, and that must fit from one line to another (Fig. 5–4).

Then there are the rarer, but more severe, problems of line locations. You may find that a line is in the wrong place on the map. If a good reflection is at radically different reflection times on intersecting lines, then you have to conclude that one of them is out of place. Or maybe one of them was just mislabeled at some stage of shooting or processing, before being entered into the computer. In either case the line obviously doesn't belong where it is shown on the map.

Moving a line to a different place on the map is a tricky step. You are likely to be wrong about which line to move and exactly where. But if you decide you have to move one in order to come up with a valid

Graphic Science Corp.

Fig. 5–4 Reflection strength at line intersection

interpretation, some systems allow you to do so. If your system doesn't have a provision for changing the locations of lines, you may have to get the change made on the survey information. That can take a long time, and call for meetings with survey people to iron out differences. A way around it, so you can get on with the interpretation, is to, as some systems permit, add times at arbitrary points on the map. Pick the section and put some reflection times from it where they would be if you had moved the line. Then delete the pick from the original position of the line. The picks in their new locations can be computer-contoured with the other data.

LOOPS

Loops are the key to interpreting 2-D areas. Pick a horizon on one line until you reach an intersection with another line. Where the lines cross, match one line with another, and pick a reflection from one to the other. Continue until the picks get back to where they started, meeting the first line at an intersection. That working around to the start is called picking a loop. If the pick meets itself back at the starting pick, on the same part of the same reflection, the loop is tied. If it doesn't, you've made a mistake somewhere on the loop. If you fail to tie the loop, you can change the pick somewhere on the loop where the reflection looks doubtful. Then if the change makes the pick tie, all you can do is assume that change was correct.

Because 2-D areas are interpreted mostly in loops, a 2-D interactive interpreting system must display loops easily and quickly. A loop can be shown on the screen opened out as one straight line, with overlap on one end (Fig. 5–5). A pick made on the overlapping part of one end of this "line" also automatically appears on the repeat of that part at the other end. When you pick it once at either end, the pick automatically appears both places. So when you pick to the end of the loop, you meet your pick from the start of the loop, on the screen. If you displayed only the loop, without overlap, you would have to move the cursor to the pick at one end of the loop, read the reflection time from the trace-time-amplitude display that gives the cursor position, and then put the cursor on that time at the other end, to see if your picking had tied.

If you have tried to splice paper sections together to form a loop, you know the frustration of having one or two of the lines going the wrong way, so you can't put it together without somehow turning them around. You either have a new playback made with the traces in the opposite order, or if you are in a rush, just make some kind of reversed print of the section you have. The reversed print unavoidably reverses the peaks and troughs, so you have to remember that part of it is reversed. An interactive system just arranges the traces so the loop forms one straight line. It puts the traces in the order that makes the display continuous.

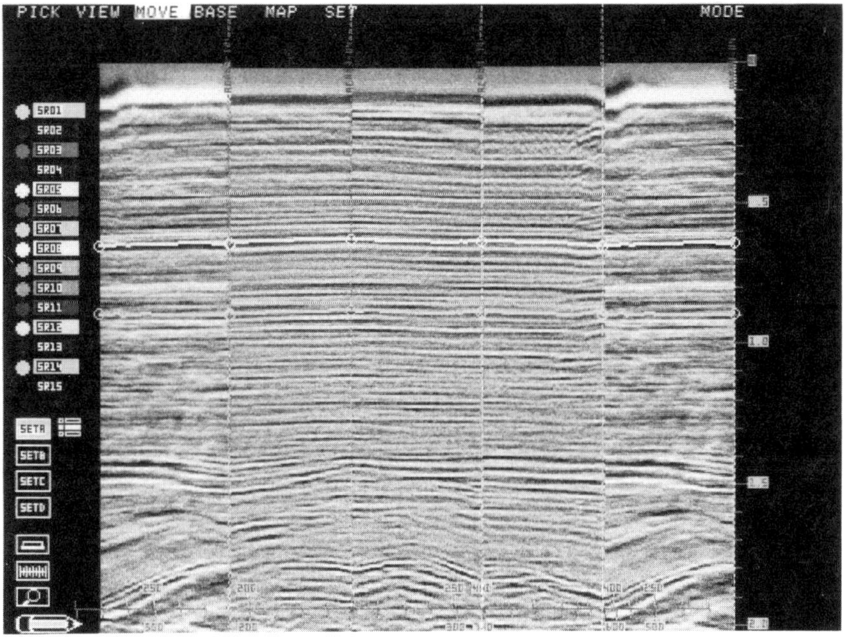

Graphic Science Corp.

Fig. 5–5 Loop straightened out

Interactive interpretation systems allow you to select a loop for display in several ways. The way that is most like getting out the paper sections is to mark, on a map display, the two ends of the part of a line you want, then the part of the next line, etc. An easier way that some systems have is to give a command for forming a loop, put the cursor somewhere inside the map image of the loop, click, and the whole loop comes on the screen.

Some systems store the loop so the next time you need it, all you have to do is call it up.

MIGRATION

The information obtained from a 2-D line is from more or less below the line. If there is dip in directions other than along the line, and there almost always is, then seismic energy received at the line is from subsurface points that are off to the side of the line. The dip on different horizons and at different places varies, so the information represents a sort of wavy surface. But since data is normally plotted on a flat paper section or displayed on a screen, it gives the impression of being two-dimensional, a vertical plane of data directly below the line. All of that is true of any seismic

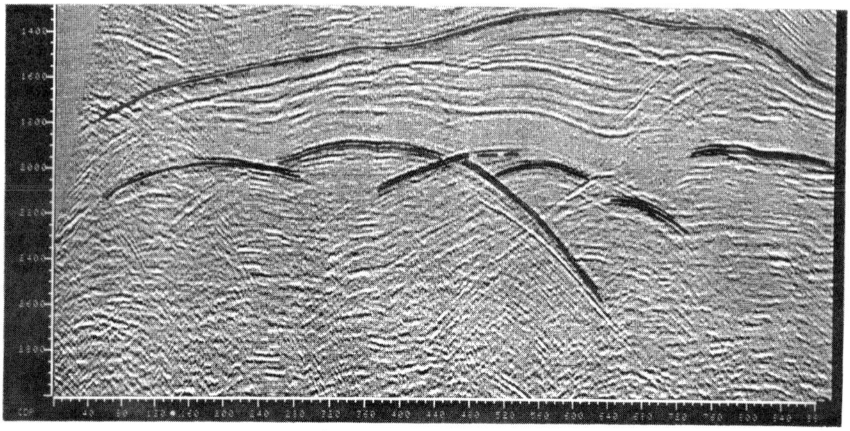

a Unmigrated

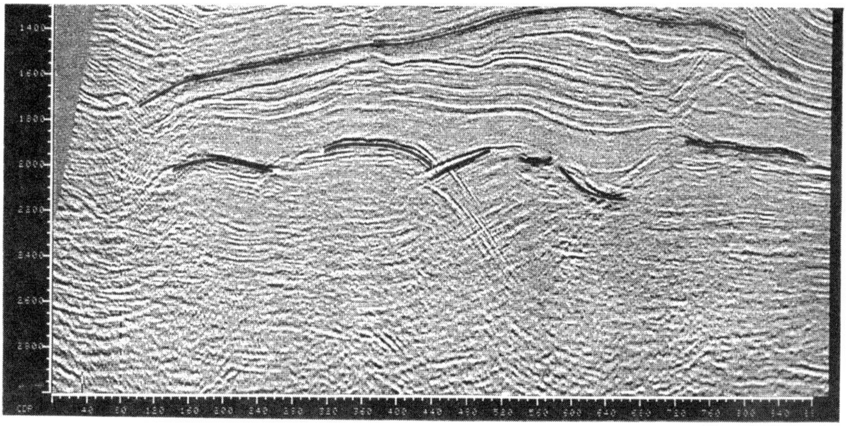

b Migrated

Petrosystems

Fig. 5–6 Same picks on stacked and migrated sections (Picks enhanced for black and white reproduction—see, not enhanced, in color on cover)

line, either 2-D or 3-D, but it leads to a particular problem in the migration of 2-D data. Reflections are migrated along the line, as though the reflections really were all straight below the surface position of the line. The migration out to the side of the line can't be made. In the extremes, a line that extends up and down dip is migrated correctly in moving the data along the line, but a line along strike isn't moved at all. The strike line should have been migrated updip, away from the line, but the 2-D shooting doesn't provide the information to move it in that direction.

Migrated 2-D lines will not tie exactly at intersections. But they will tie fairly well if there is very little dip. They will also tie reasonably well if the lines are shot at an angle nearly midway between strike and dip, so there are no dip or strike lines, and the lines in both directions have about the same incorrect placing of reflections. For interactive interpretation of 2-D data, it is best to have both migrated and unmigrated data loaded in the computer.

There is a long-standing problem of comparing migrated sections with the same sections unmigrated. But migrated sections don't tie at intersections, because dip lines are more completely migrated than strike lines. So with paper sections, you are in a quandary as to whether you should interpret the individually-more-correct sections or the ones that can be correctly related to one another.

Some interactive systems offer assistance with this 2-D problem. Both migrated and unmigrated sections can be displayed at one time, on different screens, or in different windows. When the data is loaded in memory at the start of the project, each section is loaded in each of the two forms, migrated and unmigrated. The migration velocity function is also stored. The program contains a migration formula, so it can migrate segments of picks. The formula can also be used in reverse, to de-migrate a segment that is already in a migrated position. When you pick a reflection on the stacked (unmigrated) section, it is displayed on that section, and is also migrated and displayed in its proper place on the migrated section. Similarly, if you pick on the migrated section, your pick is de-migrated and shown on the stacked section. Thus, whichever section you pick, your horizon is displayed on both (Fig. 5–6). That enables you to pick on the more-correct migrated sections, and see how the picks tie on the stacked sections. Or you can pick at the intersections on the stacked sections that relate to each other correctly, and see where your picks fall on the migrated sections.

Another solution can be used if a 2-D area has dense coverage—interpolating a regular grid of data from the irregular 2-D lines, so 3-D migration can be applied. I'll discuss that in the segment on interpolation.

THE 2-D OVERVIEW

The special characteristics of a 2-D area determine how you will get your first overall look to see what you are confronted with. For different ways in which areas are shot, the overviews will vary somewhat.

An area that has been shot in a regular grid is uniform enough for you to get an orderly first look. There are some simple, basic things you won't know about the data yet, and it's better to learn them at the start than to be surprised by them in the midst of your interpretation. What is the general structure in the area? Are there any possible prospects that are easy to see? Are all the reflections good or are there parts of the area where surface or near-surface conditions make for poor data? Does one of the good reflections pinch out, so it doesn't even exist in part of the area?

You might first call up the dip lines in succession, just glancing at them to get a general impression at this stage. To see the whole lines, you may need to use the closest trace spacing the system will display, and you may also need to decimate some traces. This is a rough overview, so you don't need to be concerned about possible aliasing of small features by over-decimation. To get some ideas of possible surprises, call up each section in turn. Since these lines are in a neat grid, you can step through the area from line to line almost as though it was a 3-D area.

After the dip lines, you can take the same sort of look at the strike lines. They won't show you regional dip, but may give you additional information on local features. Even if the strike lines are considerably longer or shorter than the dip lines, it would be sensible to view them at the same scale as you used for the dip lines. Some of the short lines may look like skinny vertical strips on the screen, and the long ones may not completely fit on it at even the compressed horizontal scale you are using, so you may have to scroll to see them completely. But keeping the compression the same, you are able to compare sizes of features from line to line. Of course, the strike lines aren't truly strike lines any more than the dip lines run exactly down dip. These more-or-less strike lines may show structure on them too. To get a line completely along strike, a curving line would have to be shot along a contour line. Contours are strike lines, aren't they, bending where the strike changes?

Then, to get a closer look you could select a few representative lines and take more time looking at them. You could display them with the closest trace spacing, but without decimation, and if necessary look at each section in stages. You might even make some picks, not of any special horizon, just to get the feel of the area.

If the area is not so uniform but, like many 2-D areas, is composed of lines crisscrossing in no particular pattern, you won't be able to go through it in such a straightforward way. You may find it best to start with a line that goes through the center of the area, and then branch out to other lines more or less parallel with it, and finally to all the rest of the lines. Not having seen them in an orderly manner, you may not have a firm idea of the features of the area. To help with your visualization, you can then look a few of the lines over carefully until the main features of the area—structure, reflection quality, etc.—are familiar to you.

A still different situation occurs when you have a long regional line, maybe across an entire sedimentary basin. Then again you want a high degree of compression. The compressed version is your overview; you'll probably make actual picks on the section in a more stretched-out form. But the compressed version shows you directions of dip and subtle features made prominent. You will probably also occasionally use the compressed display later as you work the full-scale section.

After your overview, there may still be surprises when you work the area in detail, surely will, but probably not really big ones. You've seen the dip and some things about the structure. The overview has given you a good idea of the quality of the reflections.

INTERPRETING 2-D DATA

After the overview, you are ready to start interpreting the area. Here is an example of how it might go. You can establish identifications of horizons, probably with a synthetic seismogram. Insert it in a line at the location of the well it was made from, with formation tops marked on the synthetic. Decide which reflections you will pick over the area. Start picking one of them, or several at a time. At line intersections, you don't have to worry much about getting the ties exact—just keep picking on what looks like the same reflection.

You will probably pick faults as you go. At each place where a fault cuts a horizon, you can use the system's method of indicating that the break is caused by a fault. The fault breaks may show the width of the fault at that horizon (Fig. 5–7). That information will be used later, in contouring. When you have picked all the lines, a map of the horizon will show the reflection times as ribbons of color. It will also show how well you tied at intersections, either just by the contrast in ribbon colors where they cross, or by some special symbols. There are some smoothing techniques to remove the mis-ties, but you probably won't use them yet.

The best thing is to resolve the worst mis-ties first. Those are the main problems, and your interpretation will gain the most from solving them. Call up sections joined at their intersections, and try to pick correctly across. After getting the misses down to within a few milliseconds, you can have the times contoured, maybe with a "quick-and-dirty" contouring system that gives you contours that are composed of straight-line segments, so you can see how your interpretation is shaping up. Now you may want to use a smoothing function to make the times match exactly at intersections, so you don't have two different times at one point.

After making a number of fault picks you can work on getting the faults sorted out, connecting a fault found on one line with a fault break on another. This has always been difficult in 2-D, but you can make it a little easier by displaying the sections together and looking for similarities in the fault indications on the different lines (Fig. 5–8). Once you decide that two of them represent the same fault, you can give that fault a name,

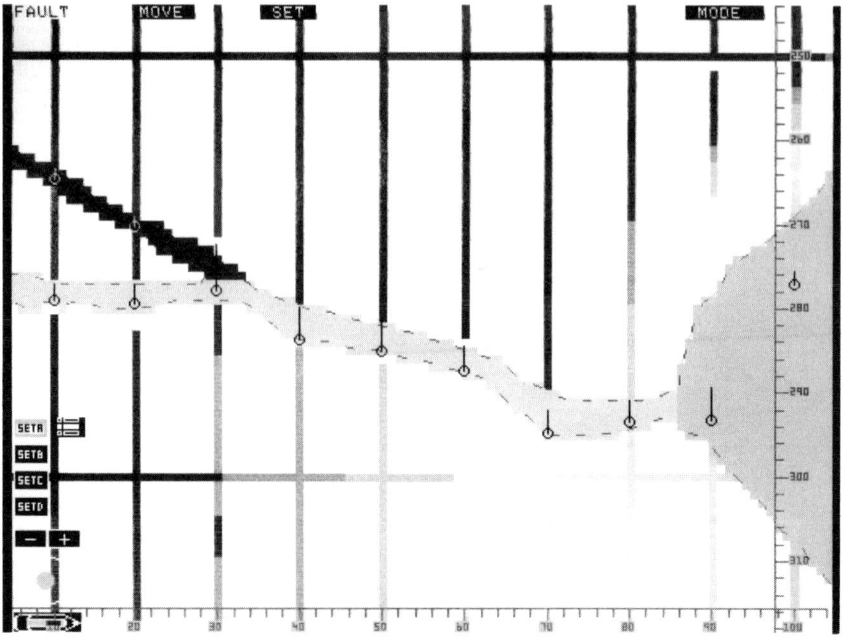

Graphic Science Corp.

Fig. 5–7 Fault breaks with heave

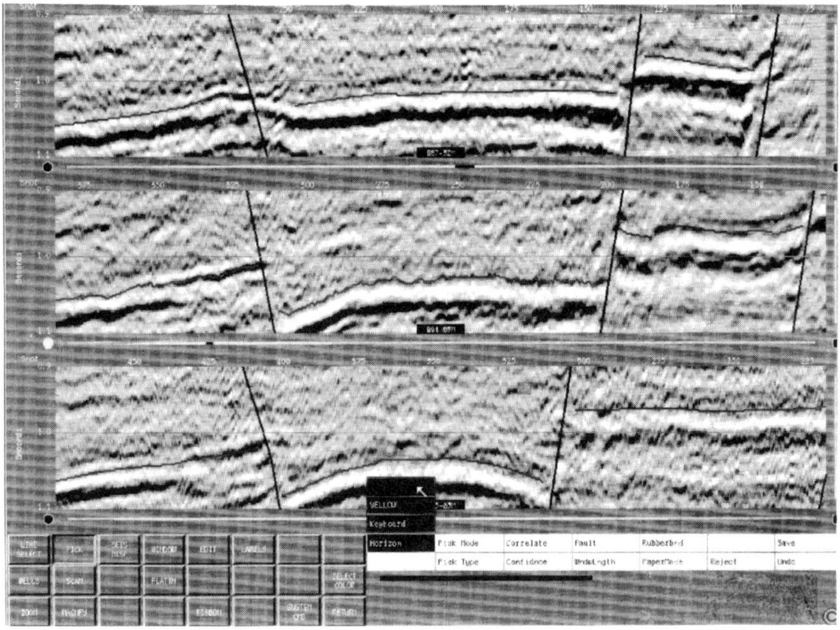

Sierra Geophysics, Inc.

Fig. 5–8 Faults on three lines compared

so it is kept in its own file, like a horizon. As you find that other fault breaks are parts of that fault, you can assign them to it. When you have the reflection picks and the faults worked out for a horizon, you can use a thorough contouring program to give you a smoothly contoured map of the horizon.

When that structural interpretation is complete, you have a basis for further investigations. You might have noticed some bright spots in doing the picking, and decide to look at them in reflection strength display. Other phenomena may show up well on frequency display, for instance. You have the data in the system, know the structure, and can go on to investigate anything interesting, with the power of interactive interpretation.

COLORIZING THE OLD SECTIONS

Old 2-D seismic data continues to provide worthwhile information. Even though much of it does not have the fine detail necessary for seismic stratigraphy and some other modern techniques, it does contain valuable structural information. This old data is useful for several reasons. It can give a regional picture. It can provide information in areas that can no longer be shot, because of development of the land or other reason. And it is much cheaper than new shooting. Old data is often already in a company's files, either from old projects, or just from the different stages of shooting one area in closer detail. And other data can be purchased non-exclusively, or it can be obtained by trading data with another company.

Putting the old data in a system gives you many advantages. Before interactive interpretation, the main factors that could let a re-interpretation be much improvement on the first one were new wells in the area and re-processing of the data. Now, with the data in your system, you can analyze the data and display it in different scales and attributes. And, like old Hollywood movies, you can see these originally black and white sections in color.

COMBINING OLD AND NEW 2-D DATA

This utilization of old 2-D data presents special problems. The problems are caused by the combining of data from different shooting projects, projects that were shot in different years by different companies with different equipment and different field parameters, and were processed differently. Those differences can make the sets of data incompatible with one another in several ways. And some sets, particularly the older data, may cause problems in data loading.

In interactive interpretation, systems designed to handle 2-D have ways to adjust for some of the compatibility problems. You call up parts of two sections, so they meet on the screen at their intersection point.

They don't correlate! What can you do? Well, they may have been played out with different polarities. Try reversing the polarity of one section. That often makes it fit the other section nicely. Each reflection has a characteristic polarity arrangement, depending on the velocity difference at the rock interface and on the polarity convention used. A peak may be at a change to a faster velocity, or a trough may be at that kind of interface. You can usually match polarities reliably.

If you decide that the polarities of two lines, or sets of lines, are different, then reversing the polarity of one set may bring them into agreement. Even if it doesn't, the polarities should be made the same. After that, bulk shifts can be used to bring times into agreement. Once the two sections correlate, you may want them to look more alike, so you can change the amplitude of one to more nearly match that of the other.

It's this simple only if you have just two sections, so you can freely change the polarity and the times of either section to fit the other. Instead of just two, though, you normally have a number of lines, intersecting in a number of places. Instead of freely adjusting any section to fit another, you have to make it also fit the other sections it crosses. That is a sort of juggling act. One feature of interactive interpretation systems that can help you in this is a map display of mis-ties. To use it, you have to first pick a horizon. For this purpose, you should choose a strong reflection, one that you can recognize easily, then pick it on each section independently of the others. Display the mis-ties, and look for a pattern in them. Maybe one line is consistently lower than all the others it crosses. Or all the sections of one shooting project may mis-tie another set by a more or less uniform amount. A pattern like that can allow you to make an adjustment that is consistent, and therefore about as near to correct as you can get.

The difference doesn't tell you which set to shift, and as long as they fit after the shift, it doesn't make much difference in the picking. Of course, if one set ties wells better than the other, then the one that fits the wells should be left unchanged. Or, if there aren't any wells in the area, one set of seismic lines that mis-ties several other sets is the one to change. And, even if there are just two sets, if one set has a lot more seismic data than the other, then it is reasonable to change the smaller set, not only because it takes less time, but also because, with less to change, there are fewer chances to make mistakes.

Once you manage to get your different vintages of data compatible, you are ready to go on with the interpretation. You should be alert to some ways the combination of projects differs from a single project, either 3-D or 2-D lines.

On the base map on screen, the different batches of shot points may be in different colors, to distinguish one set from another. The patterns of lines probably look as though in each period of shooting, lines were assigned without much attention to the location of lines in older surveys. Some lines almost duplicate others. That ignoring of old lines happens sometimes in planning a shooting program, but also some of the old sets of data may have been shot by other oil companies, and bought from data exchanges, so each set was planned without consideration of the others.

The combination calls for a special way of interpreting. When you have a large amount of new data to incorporate with old data, it is usually best, if you have the time, to use the new as the basis for your interpretation, and fit the old into it. That is the reverse of what seems natural to do, but there are several reasons for it. The new data is almost certainly better than the old because of the progress in shooting and processing since the other surveys were made. And working the new part first will also free you from the "historical" influences, the prejudice you might unavoidably have in favor of the old interpretation, and the influence of the particular arrangement of data the earlier interpretation was based on. After settling the main questions in your interpretation, you can compare it with the earlier interpretations that other people may have made, so you don't fail to use useful ideas those interpreters had.

LOADING OLD AND NEW FROM TAPE

In normal living, there are a number of forms to fill out, like income tax reports and driver's license applications. Many of them are confusing; the designers of the forms didn't think them out well. Although it is important to fill them out correctly, sometimes illegal to not fill them correctly, you may have to guess at what the designers had in mind.

Loading seismic data can be like that, using bits of evidence like a detective does. You have seismic tapes with data on them. The tapes may not have all the information needed for loading; like the government forms, they may require some guessing. It may be difficult to determine shot point numbers or CDP numbers. Maps and paper sections may have some of the missing clues. That information is vital to loading your data correctly. If it is done incorrectly, the displays on screen might look all right, but a drilling location you determine from the data won't be in the correct place.

IS IT TOO SMALL AN AREA TO BOTHER LOADING?

If you are going to interpret a small 2-D area that hasn't been loaded into a system, you may decide that it isn't worth the time and effort to load it, that it may be simpler to just work it on paper. People normally do consider that it isn't worthwhile to load an area if you can interpret it in a few days on paper. That is often a good judgment, especially if you are in a hurry to get the interpretation done in time for a lease sale or other deadline.

But there are some reasons that might justify loading even a small area, if the time is available. For one thing, the loading would make it ready for not only the interpretation you are doing now, but for all the subsequent looks at it by you and other interpreters. Also, you may be able to interpret better on an interactive system than on paper. Even if it is old shooting without the true amplitude preservation that makes many advantages of modern interpretation possible, you can have the automatically quick and correct reading and posting of picks. And your interpretation as well as the basic data will be available later for calling up on the system instead of hunting up on paper sections and maps, to make checks of expiring leases, etc. So, even for a small area, it may be better to load the data. If you don't, then the next time anyone does any interpretation on it, the decision of whether to load or not will have to be made again, and again and again until someone decides to go ahead and load it. Then you may look back at the times the area was reviewed in the old way and wish someone had decided to load it that first time. One loading takes the same amount of time whenever it is done, so the sooner it is done the more benefit the company will get in the long run.

WITHOUT TAPES

Taking data from tapes and entering it in a computer is the best way to make it available for interactive interpretation, but there are situations in which it either isn't possible or just isn't practical. The lines might have been shot before tape came into use. Or you may have sections and maps received by trade or purchase, with no tapes accompanying them. There may be old sections in your company's files, but the tapes those sections were made from may have been misplaced.

Even though companies may have the tapes in their files, they may sometimes find it easier to transcribe data from paper sections than to play it back from tape. The tapes stored are usually not final tapes, but unfiltered tapes, kept in that form so replays can be made with different filters. It can sometimes be quicker and cheaper to transcribe from paper than to dig out the tapes, filter, and load into the system.

Almost always, the data in these situations is 2-D, not 3-D. But it is possible for 3-D data to be in the same condition, for instance if the tapes for an area have been accidentally lost or destroyed, but a set of prints remains. There could even be a situation in which the tapes were far away, maybe overseas, and you had prints and wanted to do some interpreting on a system.

There are a number of ways to get information into the system without starting from the processor's tapes. Some of these methods you or a technician will be able to do, and some will need the services of specialists. The quickest way to get the results of an interpretation into digital form, but also the most superficial, is by digitizing maps.

DIGITIZE MAPS

For digitizing maps, there are two main types of digitizer. One is a graphics pad like the graphics pads on some interactive systems—a board with a fine mesh of wires under its surface. The pad for digitizing maps has to be much larger than one used just to move a cursor on a screen, so it is usually built into a drafting table, which may then be called a digitizing table. When a stylus or puck is moved around on the pad, its location is sensed by the wire mesh, and transmitted to a computer. To indicate a point, you put the crosshairs of the puck or the tip of the stylus on the point and click. A stylus feels natural, like drawing with a pen, but you may be able to be more precise with a puck, if you lean over the table so you are looking straight down at the crosshairs. If you look from an angle, the crossing point will not be where it appears to be.

The other type of digitizer is a long device that is mounted along the back edge of a drafting table. It contains a track with a perpendicular arm that runs along the track. A slider that can be moved up and down the arm has a window with crosshairs. Positioning the crosshairs over a point and pressing a button sends the location information to the computer. This type of digitizer doesn't need the grid of wires, so it can be mounted on an ordinary drafting table. It senses the position of the arm and that of the slider, to record the location of the crosshairs.

Either kind of digitizer may include a numeric keypad (a small keyboard with number keys on it) so you can enter data without turning to the computer's keyboard. The digitizer may also display what you have digitized on the computer screen, so you can see how far you have gone. Instead of the button to press for digitizing a point it may have a foot pedal, so your hands can be devoted strictly to positioning the sensor.

Digitizing is about the same with the two kinds. You tape the map down, putting it anywhere on the table, and calibrate it by clicking on three corners or other points, and keying in their coordinates. The three give the computer the scale, position, and orientation of the map. Then you can point and click on whatever shot points or other locations you are entering. On these points you don't key in numbers for location—the computer determines locations from the points' relationships to the three calibration points.

With a digitizer, map scale isn't a problem. When you mark the three points and give their coordinates, you are giving the scale. But if you have to enter data from different maps that are on different projections, then you need a program that converts from one set of coordinates to another.

The first information to digitize from a map is locations: locations of shot points, wells, survey bench marks, and any topography, culture, or lease information that might be useful. That gives you the basis for maps on which you can plot other data.

After you have the locations, you can add other information that applies to those positions, information like reflection times at the shot points and formation tops at the wells. You can type them on the keyboard, or if they are already in magnetic form, a program may be able to combine them with the location information so they can be displayed on a map. In any case, if the numbers are there, the computer can contour the data. The new contouring may provide a concept that wasn't thought of originally. You can edit the computer's new contouring to make it look more geologically reasonable. There is also a chance that you may find errors in the old contouring.

If you have contoured maps without the times or other data that the contours are based on, you can enter the contours by digitizing points along them. Then you have only the contours from the original interpretation, but they are in the system, so you can call them up, overlay them on other information, in general get more use from them than if they were only on a paper or film map filed away somewhere.

TRANSCRIBE PICKS

The next step beyond just digitizing contoured maps is digitizing picks that are already marked on paper sections. It is a good way to get more information from a backlog of interpretations into computer memory. The computer can then determine reflection times without error, and in doing so it may correct errors that were made back when the interpreting was done originally. It can post the times it reads to maps, again possibly correcting old errors. The picks can be computer contoured, with the same advantages there are from computer contouring of data taken from maps.

If you have a digitizing table available, you can digitize picks the same way you do map data, except that you will use a scale of distance in one direction and time in the other. It doesn't make any difference to digitizing equipment whether the units are in seconds or what—they are just numbers to it.

Some devices are made specifically for digitizing picks. Such a device has a long bar that you place along a trace. You move a sliding pointer on the bar to a point on the trace and press a button to record the point. It is connected to a personal computer, which can even be a laptop. To give it the scale of the trace, record two points, like 1.000 and 4.000 seconds. Then you can move the pointer to a picked reflection and press the button to record the reflection time in the computer. It isn't necessary to position the straightedge so it is exactly vertical on the section; marking the two reference times so it measures between them makes that unnecessary. This is more convenient for the specific purpose—picking—than a more general-purpose digitizing table, on which the section must be kept fixed, and can't be moved to more comfortable positions.

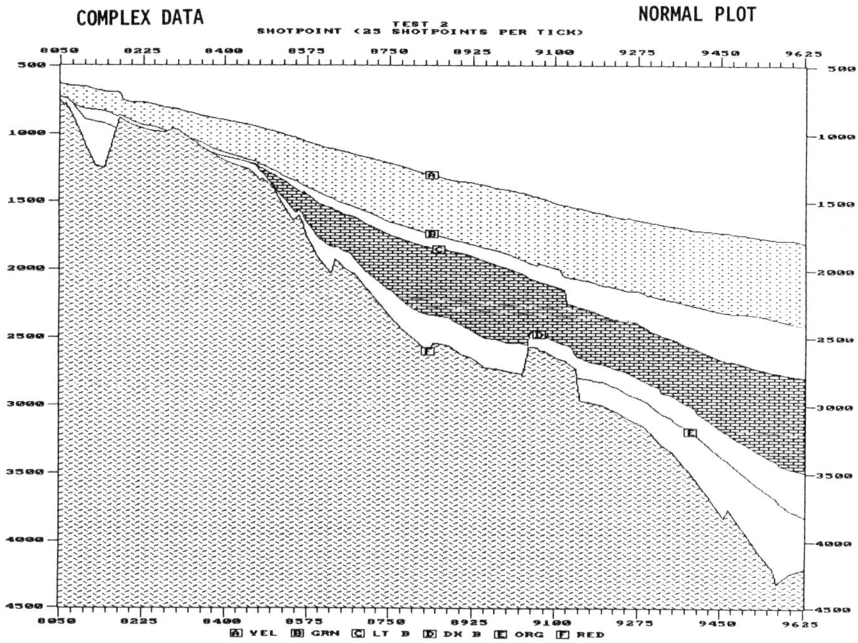

a Picks from section

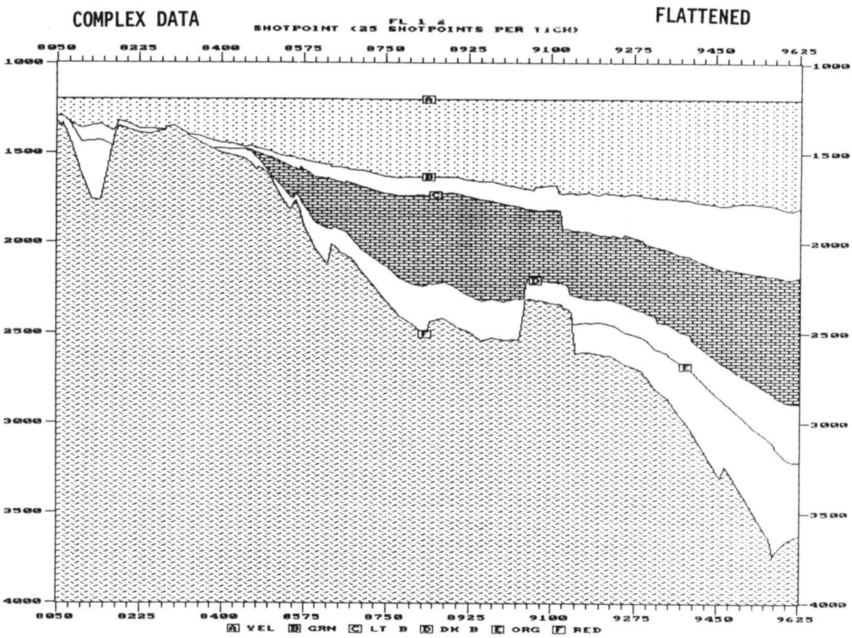

b Horizon flattened

Digi-Rule Inc.

Fig. 5–9 Printouts from digitizer

With the picks digitized, you can print them out as a list, or you may be able to plot them as a cross section (Fig. 5–9), (depending on the particular device's compromise between compactness and function) on a plain printer, even a small portable one. And of course, the digitized data can be copied to a more powerful system, if you have one available. The equipment, including laptop computer and portable printer, can be carried easily. The system can be taken to other offices and used wherever electrical power is available.

Using any of these ways to digitize picks, you can digitize picks on paper sections, either an old interpretation or one you are currently making. You again have the advantages of automatic timing. For new interpretations, since it isn't convenient to fold sections at intersections for correlating while digitizing, it may be better to just pick the whole area in the old way, then digitize your picks, and let the computer do the legwork.

SCAN SECTIONS

Without tapes, when you digitize the picks, you have the advantage of timing and posting. Another way to have this and other advantages is by scanning the paper section and storing it as an image. You can call the image up on the screen with all the paper section's header and labeling (Fig. 5–10), click on some calibrating points, and then pick. You can't manipulate the traces or display them in different forms, and you can't use automatic picking, because the traces are not stored individually, but are part of the image. You can pick by rubber band or freehand, erase, mark terminations against faults, zoom, and change proportions (by stretching and squeezing the image). You can put parts of two lines together at their intersection, or make a composite line. As with paper sections, a loop put together as one line will have some of the individual lines going the wrong direction. To get around that problem, the images of those lines can be reversed, like looking at a transparent print from the wrong side. The peaks go to the left instead of to the right, but you can pick around the loop. Your picks can be stored so you can display any horizons, have the system contour them, calculate differences between them.

TRANSCRIBE TRACES

Digitizing picks allows you to use automatic timing, but if you can get the basic data by digitizing the traces and putting them into memory, the interactive system can do automatic and semi-automatic picking, displaying sections in different forms, etc. One way of doing this is by moving a puck or stylus on a graphics table along a wiggle trace. That puts the information into magnetic memory. This would seem to be a task that could be automated fairly easily, so it wouldn't have to be done by hand, but there are difficulties that make that, at present, impractical. Wiggle

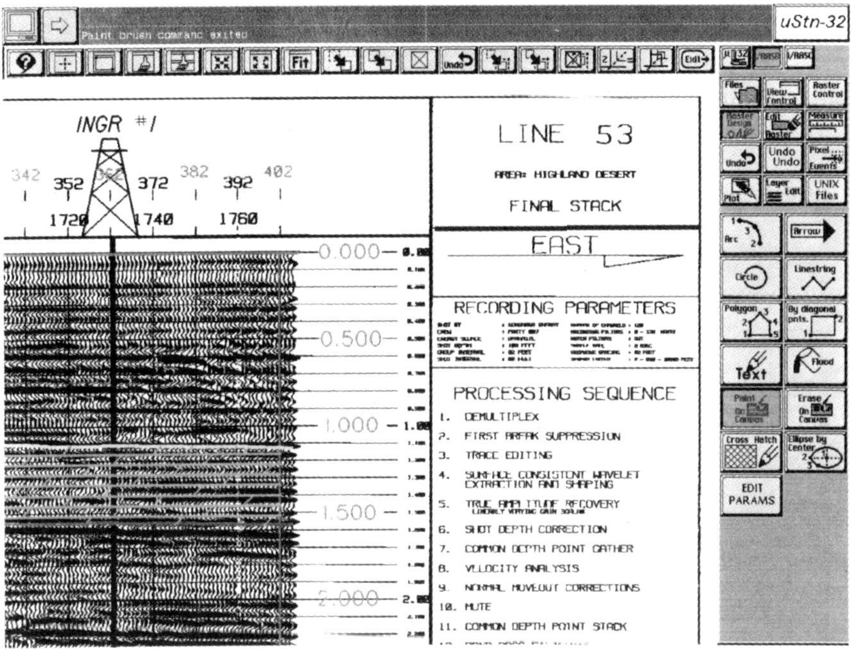

Intergraph Corporation

Fig. 5–10 Scanned section on screen

traces on paper sections are so close together that they overlap. The closeness is necessary so an interpreter can look along reflections easily. Where the overlap makes two traces cross, it is hard for a computer to stay on the same trace. And if the wiggles also have variable area fill-in, then in the overlap the computer gets lost in a mass of solid color. So capturing the individual traces is done by hand. This is normally done by a service company.

This method has the advantage of putting the entire trace in memory, so the computer can treat it like any other trace, displaying it in different colors, with different attributes, different amplitudes, different forms of trace. It has the disadvantages of being subject to human error, and of being made from a processed section so it can't be re-processed except in limited ways (for instance, frequencies still in the traces could be filtered out, but any already filtered out could not be put back in). This kind of digitizing will work only if the original that is being digitized has wiggles. It can't be made from variable-density sections or sections made of variable-area only.

A machine method that accomplishes some of the digitizing uses

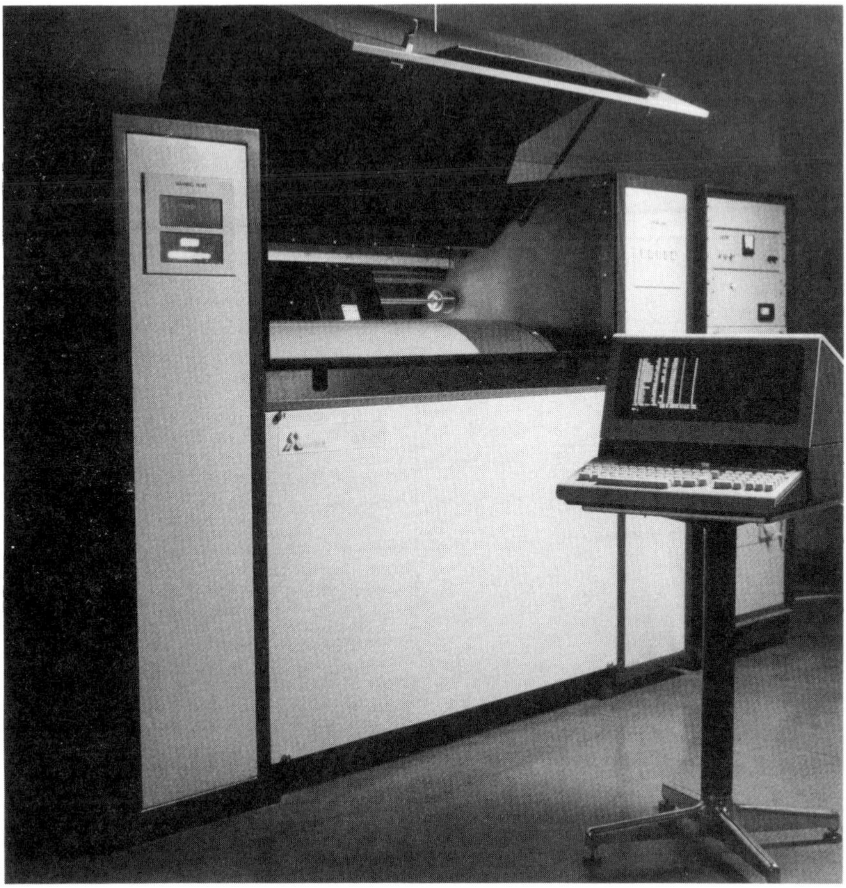

Graphic Science Corp.

Fig. 5–11 Scanner

variable-area displays, either VA only or wiggle-VA. The dark areas are sensed by a scanner (Fig. 5–11). The overlap is avoided by its sensing the part of the dark area near the central position of the trace. Thus the bases of the filled-in peaks are put into memory. From them the entire peaks can be determined. The peaks can then be manipulated like other data. The method has speed and accuracy, but it doesn't have all the data—the peaks are exact, but the troughs are missing. Filtering, though, can produce fairly good estimates of the troughs. A section much like the original is produced. This is not a one-trace-at-a-time method, but one that scans the whole section as a unit (Fig. 5–12). It is normally used to scan paper

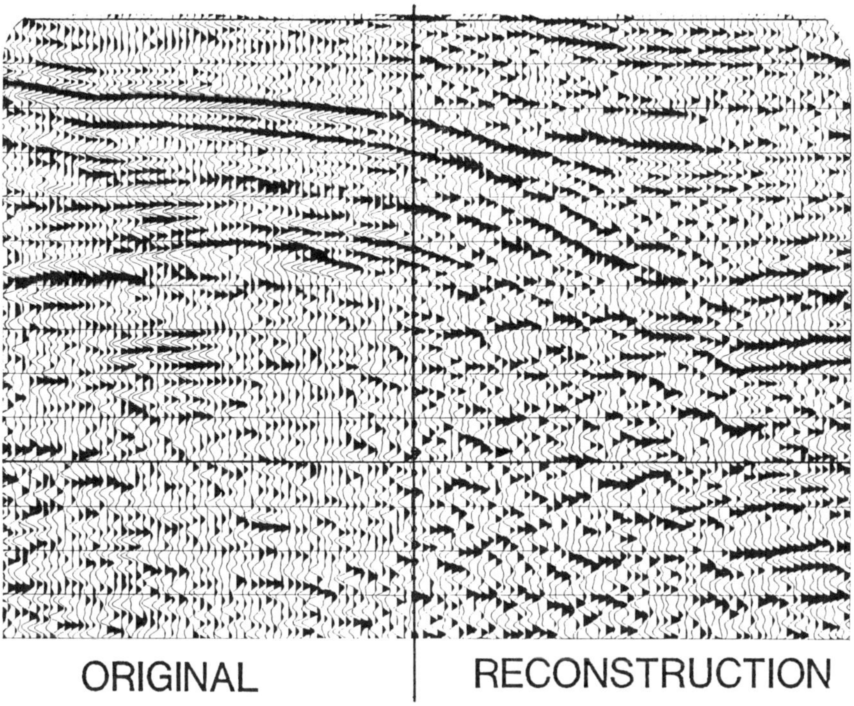

Graphic Science Corp.

Fig. 5–12 Traces from scanning and original

sections, but the digitizer is so sensitive to fine detail that it can even handle sections on the 105mm film that some companies use for storing maps and sections. The method is sometimes used to put sections on tape even though the original field tapes and perhaps tapes from later stages of processing are in the company's files. Usually the final filtered tapes are not saved, so it may be quicker and cheaper to just transcribe the peaks from paper sections. And for some purposes, like mapping regional structure, it is completely satisfactory.

6

If It's a 3-D Area

In contrast to the often piecemeal shooting of 2-D, a 3-D area is shot all together, with the surveying and recording handled so as to be consistent in all directions. This areal approach is well described by the expression "blanket coverage."

THE NATURE OF 3-D

To achieve blanket coverage, a 3-D area is shot as a set of parallel lines. The lines themselves are, like 2-D lines, shot in the repetitious manner that provides CDP stacked data. There is no need for lines to be shot in another direction to form loops. The lines shot are so close together that two nearby traces from adjacent lines look about as similar as two traces from the same line. So information about the subsurface is obtained at a network of points over an area instead of just along a line.

The actual shooting can take different forms. Offshore, the only practical way to operate is for a boat to keep moving, towing geophones and sound sources behind it. So lines are mostly shot individually, although sometimes they are shot in groups of two or more, by having the boat tow more than one set of geophones or sources, or by using two boats on parallel courses. The area is built up from parallel runs, each run producing one or more lines of data. Currents in the water divert the cable to one side or the other, so it does not follow straight behind the boat, and information from one run may overlap that from another.

In processing, the area is divided into a uniform grid of "bins" (Fig. 6–1), and all depth points, from whatever runs, that fall in the same bin are stacked together to make a single stacked trace. So, although sea conditions and other factors may scatter the depth points irregularly, the arbitrarily drawn bins are perfectly regular. This gives a degree of uniformity to 3-D data that 2-D does not have.

On land, where the shooting is not restricted to the path of a boat, and where it is subject to the slower laying out of geophones by hand, a different plan is used. The usual method is to lay out a number of parallel lines of geophone groups (although the geophones in a group may not be in a line, but extended to one or both sides), and then shoot along a line that cuts across those lines at a right angle. Binning is used in this method too, so again the data is in an arbitrarily uniform grid.

The spacing between lines is normally not as close as the spacing

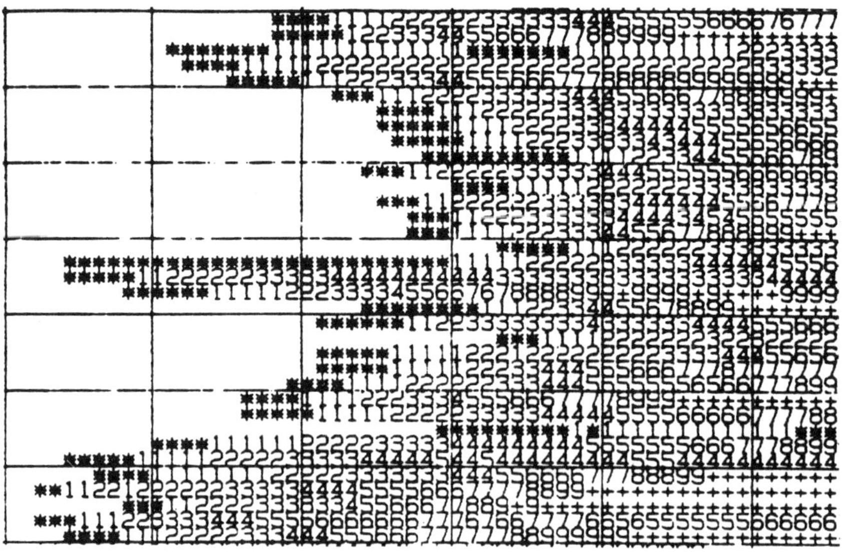

CogniSeis Development, Inc.

Fig. 6–1 Points in bins

between the traces of one line. Lines are often about twice as far apart as traces. The lines that are shot, normally run in the general direction of dip, and it is harder to see an alignment on a dipping reflection than on a flatter one. That difficulty is eased by the closer spacing of traces in the dip direction. And the difficulty of picking in the more sparcely sampled strike direction is eased by the lesser dip. So picking reflections in the general directions of strike and dip are about equal in difficulty. Any line made up from the 3-D area can be picked about as easily as one in any other direction, depending on the steepness of the dip and the line spacing that was chosen for shooting the area.

The primary idea of 3-D is to get coverage that is continuous in all directions. But there are a number of other characteristics that come from that close grid of shooting.

Since it is shot as a unit, there is—has to be—good control on the surveying, so locations and elevations (or water depths) are coherent not only along a line, but from line to line. And if some inconsistencies do somehow get through the surveyors' checking and mapping, they are not visible to the interpreter because they have been forced into consistency by the binning process. Any problems of location and binning have either been handled correctly or the imperfections have been merged in the binning before the data gets to the interpreter.

The same situation also affects near-surface corrections. They can be made coherently in all directions, and if there are any errors in them, the errors are blended in the stacking of binned data, so they are not apparent to the interpreter. That, of course, is what happens in any processing along any line, 2-D or 3-D. The phenomenon that is distinctive to 3-D is that this blending occurs not only within a line, but also from line to line.

Determination of velocity also can be made consistent in all directions, by analyzing a local area rather than just a local distance along a line.

A major result of the blanket coverage is its effect on migration. Migrating 2-D data is flawed in that the migration can occur only along lines, so if, as is usually the case, the dip is not exactly along the line, the migrating should, but can't, move the depth points off to the sides of the line. But 3-D data includes information from other lines to the sides of that line, and that information is used in migration, so 3-D data is much more correctly migrated than 2-D.

The effects of the two kinds of migration are shown in Figures 6–2 and 6–3. Figure 6–2a is a part of a stacked seismic section, that is, stacked but not migrated. It has some steep dips and some diffractions. In Figure 6–2b, it has been 2-D migrated. The dipping reflections are more coherent and the diffractions are removed. Figure 6–2c shows it after 3-D migration, with the dipping reflections still more continuous. Each migration improved the data. A greater change can be seen on a crossline in Figure 6–3. Again,

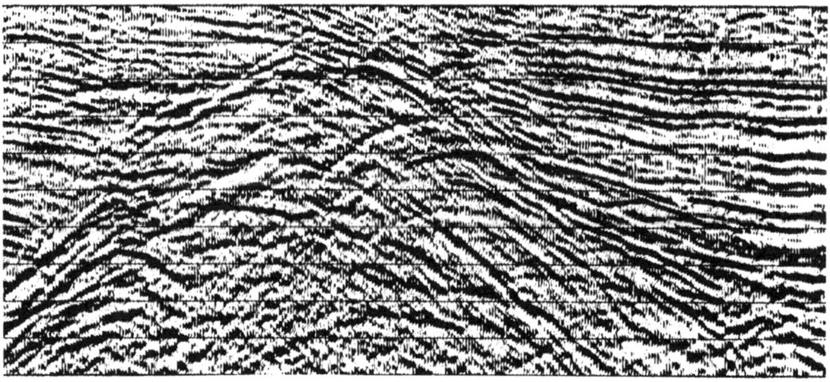

a Unmigrated

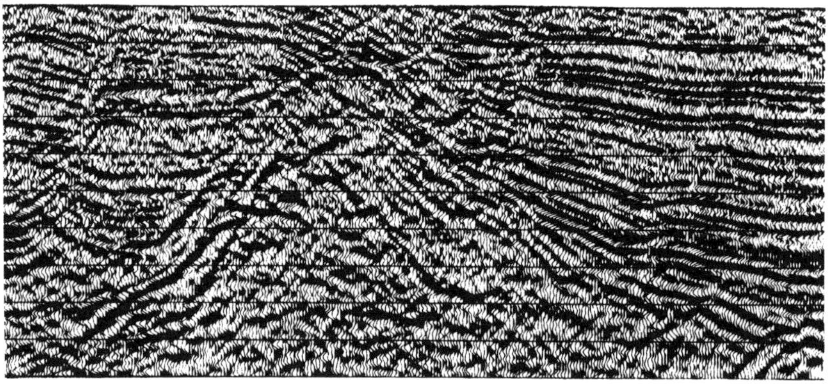

b 2-D migrated

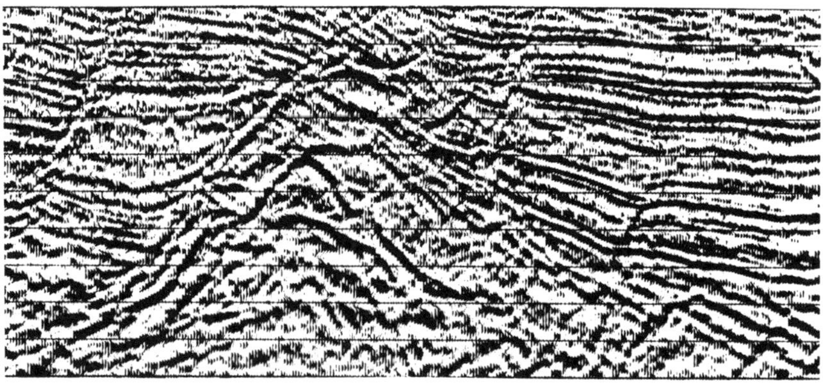

c 3-D migrated

CogniSeis Development, Inc.

Fig. 6–2 Effect of migration on inline

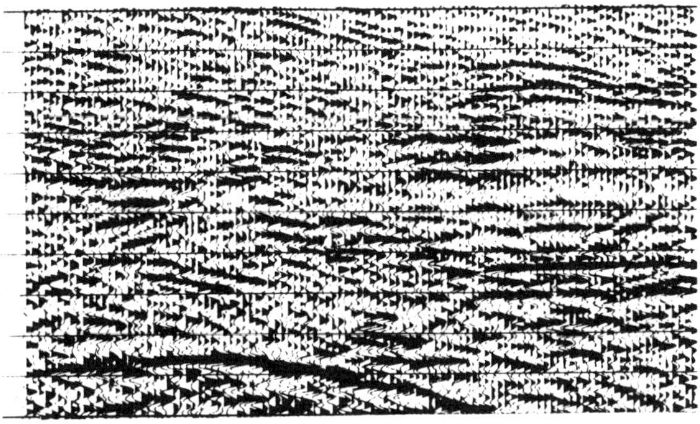

a Unmigrated

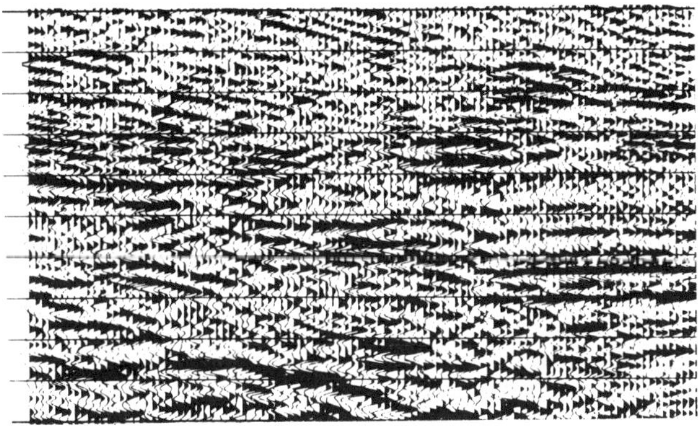

b 2-D migrated

c 3-D migrated

CogniSeis Development, Inc.

Fig. 6–3 Effect of migration on xline

the a, b, and c parts of the figure are the section stacked, 2-D migrated, and 3-D migrated. Notice that it has a strong reflection without much dip where there is steep dip on the inline. The 2-D migration improves the reflection. But on the 3-D migrated section that reflection is missing! Of course it is. With the steep dip on the inline, this reflection is from a point updip, so it isn't on the crossline.

Summing up and also comparing 3-D with the same points made in "The Nature of 2-D", Chapter 5:

- **Coverage.** The close lines of 3-D not only allow lines in all directions to be viewed, but also permit horizontal displays of the information.
- **Line Length.** A 3-D area is a coherent unit, and does not have the extremely long lines that can be found in 2-D shooting, so it is usually possible to see an entire 3-D line on the screen at one time.
- **Line Ties.** A 3-D area is shot all together, with the information gathered as a unit, so there are not the problems tieing lines at intersections that there are in a 2-D area.
- **Migration.** In 3-D the data is migrated in three dimensions, without the cross-dip problem of 2-D.

3-D DISPLAYS

Most of the discussion of displays of sections up to this point applies to 3-D data as well as 2-D. The lines of 3-D data are displayed as sections in the same ways 2-D lines are, except for 3-D not having the problems of long lines and line ties. Even loops can be formed by selecting appropriate parts of lines in 3-D surveys.

SECTIONS IN ALL DIRECTIONS

The data for a 3-D survey can be thought of as a block of samples, recorded digitally as numbers representing amplitudes of seismic energy. A sequence of samples, usually one every four milliseconds, makes up a trace, and a number of traces make up a line. All the lines side by side make up the data for the area. So the data forms a solid mass composed of lines composed of traces composed of samples. The sample is the unit of data, just a number written in the binary code of ones and zeros. Each of these samples is kept in memory in a form so it can be accessed separately from all the other samples.

The lines that are shot in 3-D are inlines, and their sections are inline sections. They are so close together that it is easy to correlate one line with the next. If you were interpreting them on paper, you could fold a section anywhere, lay it over the corresponding point on another section, and correlate the two. Doing that all along the two sections would allow you to pick the second in agreement with the first. But that would take a great deal of time, and you'd be kind of picking blind, without much idea how your interpretation was shaping up. Interactive interpretation gives you various views of the area and of your picks, so you can tell what you've done and where it's leading. This is done mostly by producing other displays than just inlines. For one thing, other sections can be made up from the traces of the inlines.

The simplest other section is a crossline (xline). A 2-D seismic line is composed of a series of traces representing points so close together that correlating from trace to trace is just a matter of noticing alignments of reflected energy on a section. A line that is shot in a 3-D program is the same. But in a 3-D area, with the different lines so close together, correlation from one line to another is similarly easy. A new line can be made up of one trace from each of the inlines that were actually shot, to make a crossline perpendicular to them. And on the new line too, correlation from trace to trace is accomplished by just noticing alignments. A whole set of xline sections can be made up, using all the traces of all the inlines.

Or instead of a line at right angles to the inlines, a diagonal line can be made up by taking a trace from an inline, then a trace from a bit farther along on the next inline, etc. So an xline might be made from Trace 45 of Line 1, Trace 45 of Line 2, etc. But a diagonal line could be made from Trace 45 of Line 1, Trace 46 of Line 2, Trace 47 of Line 3, etc. In this way a section can be made anywhere in the area, going in any direction.

That makes a line at an angle to the inlines, but it is made up of traces from a stair-stepping line. To make a smoother diagonal line, a system may use interpolation to form each trace from the traces surrounding it. That way, the diagonal line is made up of traces calculated to be like those that would be found shooting that line. Lines can be made up that cut faults perpendicularly, or extend in the direction of dip of any horizon or feature.

And since a line can be made in any direction, then such lines can be connected to make a line that changes direction. The main use for such a zigzagging line is to connect the wells in an area with straight segments from one well to the next. Another use is to form a line that avoids some poor-reflection area, so the line can be picked reliably, and the poorer parts of the area picked later by branching out from it. Or a composite line can be made to cut through several features so you can compare them (Fig. 6–4).

These lines in different directions can be made on an interactive system. You may be able to form an xline or other line at any time, and work with it then. Or you may find it more efficient to make a number of lines when you load the seismic data, or at the start of working the area, and

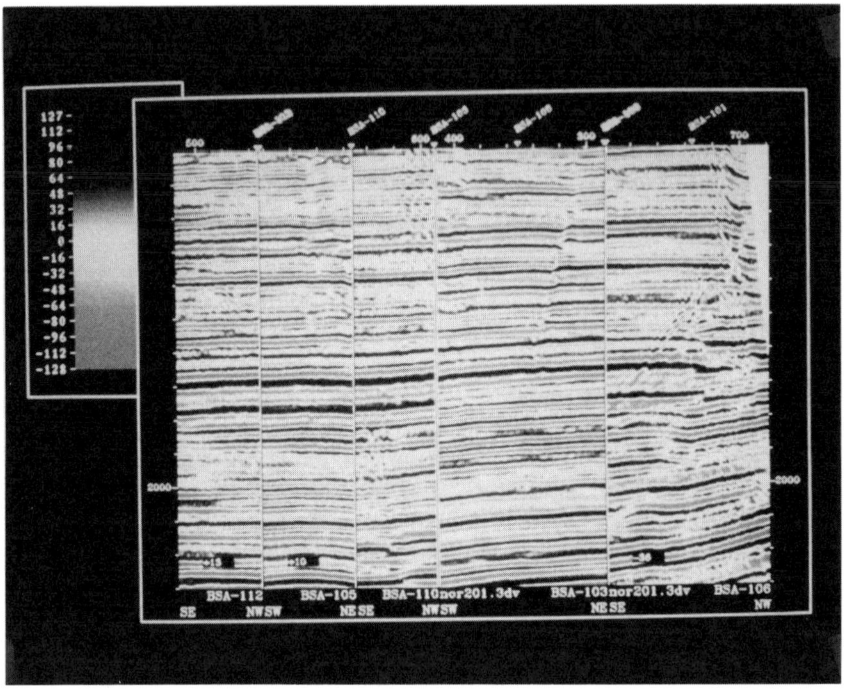

Landmark Graphics Corporation

Fig. 6–4 Zigzag line

store them on disk. Usually people do make a set of xlines, not every one possible, but enough to provide good coverage of the area.

Depending on the system, in one way or another a display can be made of lines that do not meet, for instance, several parallel lines. A system may be able to show them with space between them to show that they are separate. Or you may have to see them touching, so they look as though they did intersect. Or they may be shown one above the other. Or a system that uses windows may allow you to see a number of pieces of sections anywhere on the screen, even overlapping. If they overlap, commands let you move one window that appears partly hidden by another, and show it "on top" of the other.

TIME SLICES

A time slice is like a section, but horizontal. Time slices are sometimes called horizontal sections. That is a technically correct-enough term, but using it calls for always specifying whether a section is vertical or horizontal. To avoid clumsy expressions or confusion, I'll continue to call the vertical ones sections and the horizontal ones time slices. A section then is a display of all the data samples on a vertical plane. A time slice is a display of all the data samples on a horizontal plane. So the section has all the times on all the traces of one line, but a slice has all the traces of all the lines at just one time.

A slice looks like a contoured map of a horizon, with the reflections looking like contours (Fig. 6–5). But the curved lines are not the different contours of one horizon. Rather, one curved line on it is a time contour of one horizon, and the next curved line is also a contour, but of another horizon, and so on. That amounts to a collection of contours of different but nearby horizons, like outcrops of successive rock layers exposed on flat ground.

To get a true contour map from time slices, you need to take one contour from one slice, and the next from another, one that is one contour interval deeper or shallower. So if your contour interval is 20 ms, if you call up one slice and trace a contour on it, then display one that is 20 ms deeper, and trace a contour of the same horizon from it, etc., you will have made a map contoured in time, with a 20 ms contour interval. For this to work, the contour interval has to be a multiple of the sample rate. For instance, with data gathered every 4 ms, you couldn't use a 10 ms contour interval. A time slice is a collection of samples, and with 4 ms sampling, you can have a slice just every four milliseconds. Ten milliseconds after one slice would be between slices.

How do you know which curved reflection on one slice is the same horizon as one on the next? It may be easy to recognize, especially if there isn't much dip. But if it isn't easy, you can follow the horizon on a vertical section that intersects the slices to get from one slice to another. Contouring on time slices isn't as easy as it sounds, though, because the reflections appear much wider on a slice than they do on a section. Although the velocity interfaces that cause reflections are nearly perpendicular to a section, they are often almost parallel to a slice, so they appear broader on it. You are left with having to draw a fine line through a broad band of color. This can be alleviated by choosing colors for the display to include a narrow band of a distinctive color at some part of the reflection, like a zero crossing, and picking that band.

HORIZON SLICES

A time slice is made from samples at the same reflection time. If you flatten a picked horizon, that horizon is then all at one reflection time. If your system can flatten a horizon on all the lines, not just on one at a time, you can flatten the horizon and make a time slice at that reflection time. All the samples of the slice are from that flattened horizon. That special kind of time slice is called a horizon slice (Fig. 6–6). The broad-

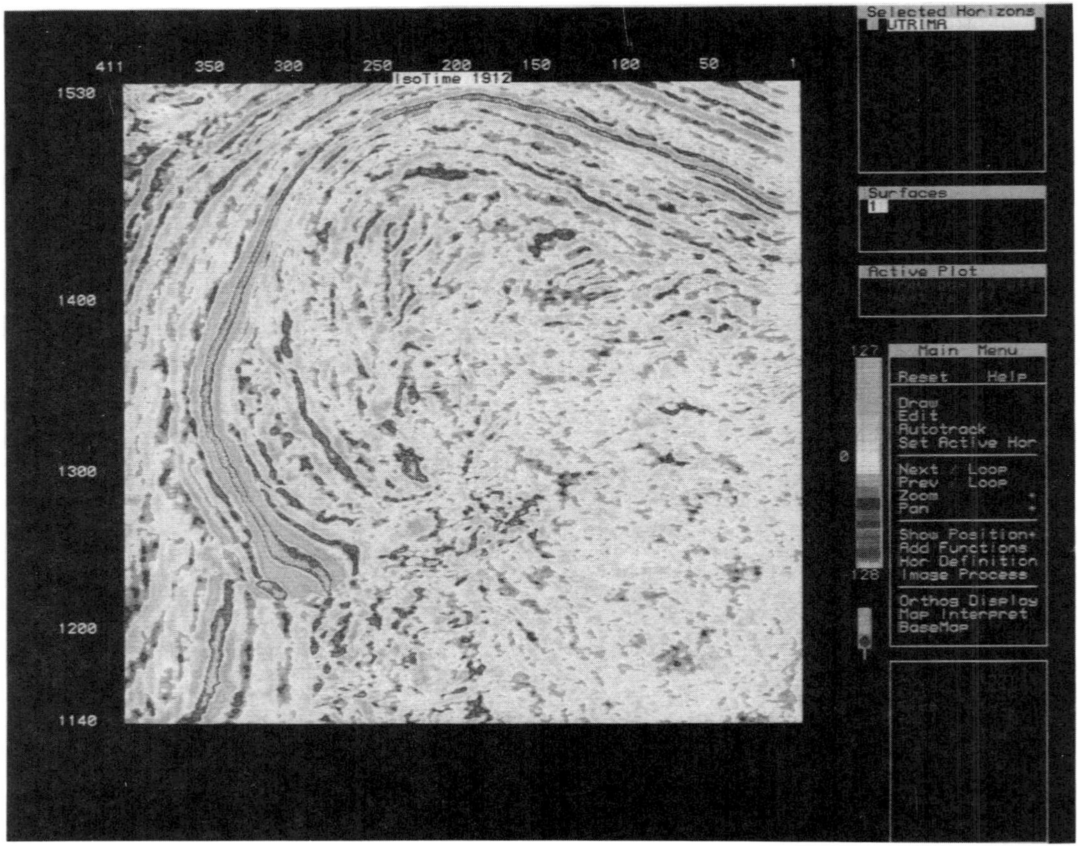

Geco Geophysical Company, Inc.

Fig. 6–5 Time slice

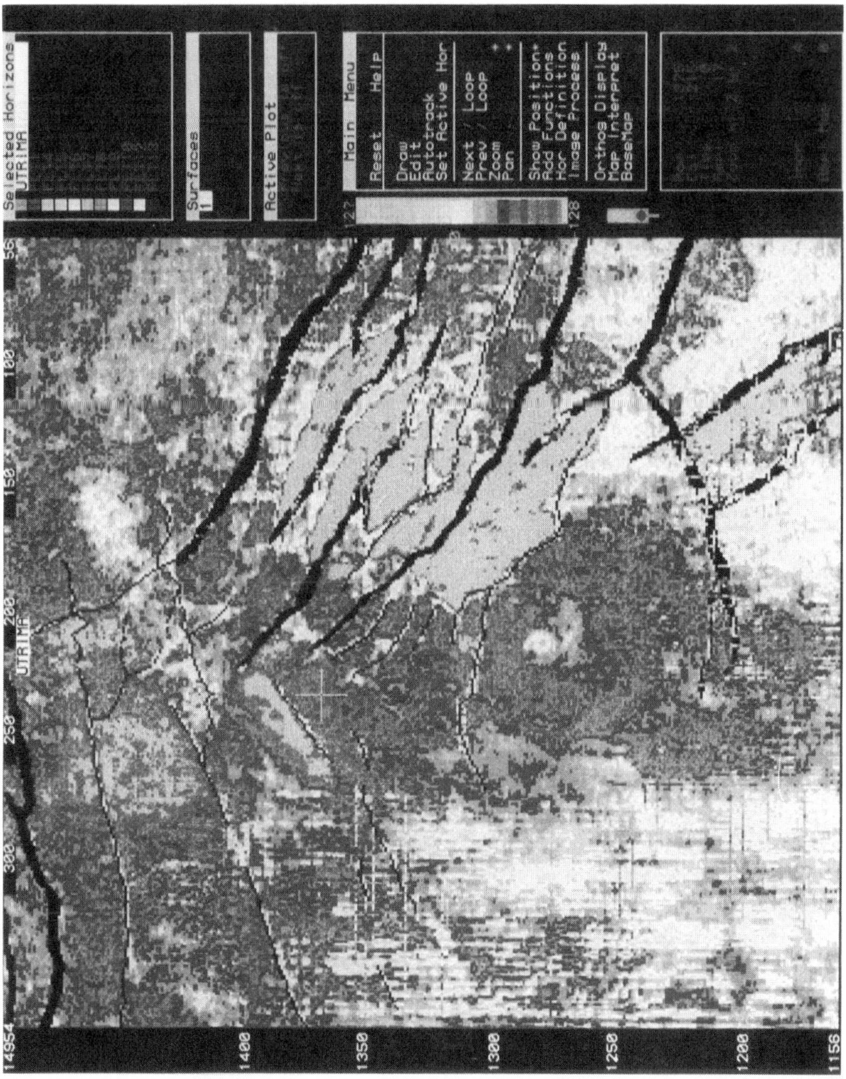

Geco Geophysical Company, Inc.

Fig. 6–6 Horizon slice

reflection phenomenon is carried to an extreme in this case—the reflection is so wide that it occupies the whole display, leaving no room for any part of any other reflection. This slice isn't just nearly parallel to the reflection, it is exactly parallel to it. The color assigned to the peak or trough is all over the display. But a display that looked the same all over the screen wouldn't be very informative. So a color pattern (maybe just shades of one color) is assigned. It wouldn't help for the colors to show reflection time; you could have displayed reflection times without bothering to flatten. So you vary the color according to some other characteristic, usually amplitude. Then it isn't all uniformly colored, and any variations in the amplitude, or whatever other attribute is displayed, appear as variations in the colors or in the darkness of the single color.

A horizon slice can also be formed without flattening, by displaying the samples at the times of the picked horizon at all traces. The result is exactly the same; you've just obtained it in a different way.

A horizon slice is a powerful interpretive tool. Differences in amplitude or other attribute may make it easy to see some feature like a pattern of drainage that cuts through the formation. All this, of course, depends on the horizon's having been picked correctly before it was flattened. Imagine how you'd feel if you flattened a horizon, found a dramatic amplitude change cutting across it, showed it to the people you work with, discussed its significance, and then discovered that you'd jumped a leg in picking it, that the change in color occurred because you had picked one peak over part of the area and another over the rest, and the first one had a higher amplitude than the other.

COMBINATION DISPLAYS

Using a section to pick a horizon from one time slice to another, or the other way around, using a slice to pick from one section to another, calls for having the two kinds of display on the screen at once. There are a number of ways to show a combination of types of display.

One handy display for comparing different views of data combines a time slice and a vertical section, joined along their line of intersection, and both appearing flat on the screen (Fig. 6–7). It is the simplest combination, as though paper copies had been taped together and laid on a desk. Such a display on the screen is like a wall with a hinged shelf. A display with the section above the slice is like the view when the shelf hangs down, so you see the upper part of the wall and the upper surface of the shelf. Showing the slice above the section is like having the shelf raised against the wall, so you see the under side of the shelf and the lower part of the wall. The combination can be switched between one form and the other, so you can see how a reflection on the slice extends into the upper and lower parts of the section, like raising and lowering the shelf. The whole

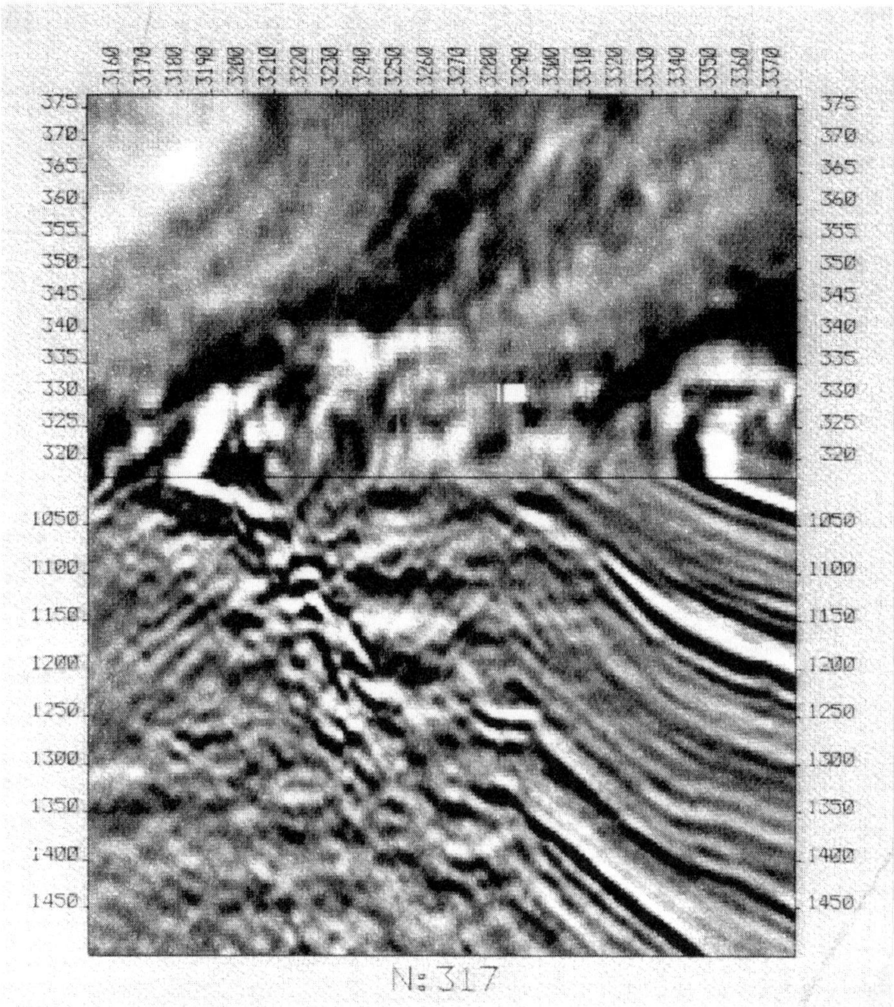

Western Geophysical Division, Western Atlas International, Inc.

Fig. 6–7 Slice and section

display can also be reversed, like going around to the other side of the wall and finding that there is another hinged shelf on that side, so you can view it up or down. Then you can see how the reflection extends on the other part of the time slice. Some people call such a display a chair, although the term also has two other meanings that we'll get to shortly.

The other types of display are more complicated. They are used mostly to analyze problems and to show features to other people.

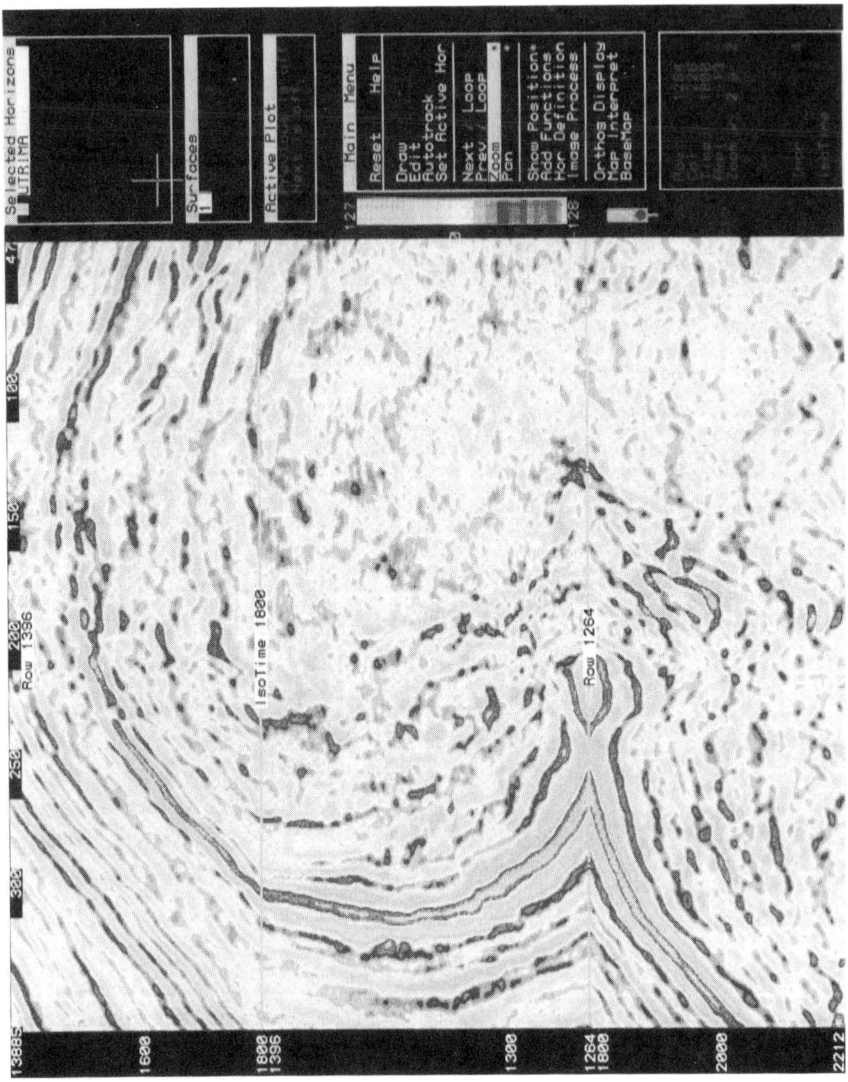

Geco Geophysical Company, Inc.

Fig. 6–8 Slice and two sections

A display can have one slice and two sections, the front one being the part of a section that is deeper than the slice, and the back one the shallower part of a different section (Fig. 6–8). This display too is free of distortion, with the three parts appearing flat on the screen. This is another type of display that is called a chair.

More information can be displayed by showing a slice with four sections around it, as though taped to its edges (Fig. 6–9). The display forms a sort of plus sign, and the slice and sections are not distorted, but the sections, which actually join to form a loop around the slice, have big open rectangular areas between them in the display. With this kind of display, it is easy to pick from the slice to a section, and from a section to the slice, but not from one section to another.

Those displays are all flat, without distortion. But there are also semi-perspective displays, looking something like photographs of solid objects.

A display can be made of combinations of sections shown in perspective like a model made of sheets of cardboard. It can form a fence diagram or something more complex, like adding a time slice to a fence diagram to give it more continuity (Fig. 6–10).

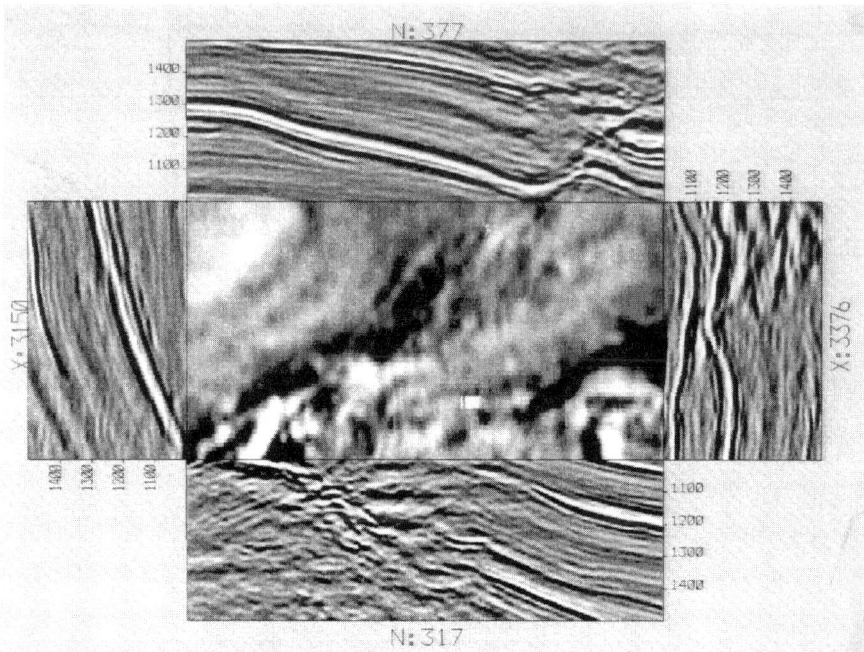

Western Geophysical Division, Western Atlas International, Inc.

Fig. 6–9 Slice and four sections

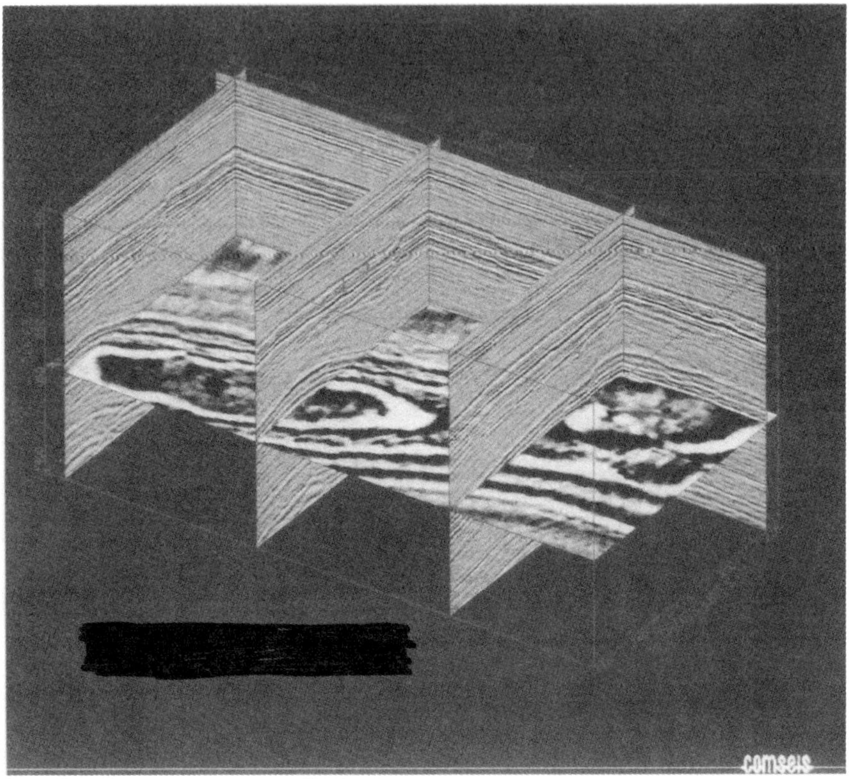

PRAKLA-SEISMOS AG

Fig. 6–10 Fence diagram with slice

Two sections and a slice can be joined, with the slice at the bottom, and the combination displayed as a perspective view into a corner. Or the slice can be put at the top, so the perspective view looks like the outside of a solid block. Although the sides aren't necessarily square, it is called a cube (Fig. 6–11). In this kind of display, only the front of the display is undistorted. The other surfaces are in semi-perspective, so they aren't as easy to work on. Their edges are angled as in perspective, but there isn't any foreshortening. The farther parts don't appear smaller, so the edges don't get closer together at the far ends, as in true perspective. In artists' terms, they don't approach a vanishing point.

If a cube display is made with the two back sections extended above the top of the cube, some people call it a chair, the third shape I've described that is called that (Fig. 6–12). The cube with extended back sections has both the front face and the "chair back" not distorted by perspective.

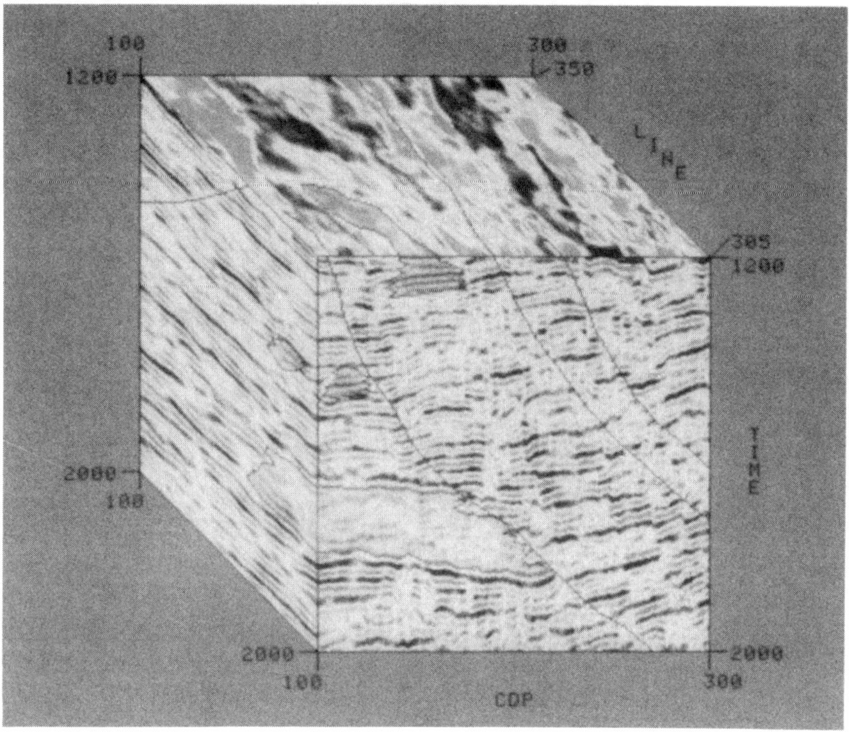

Sierra Geophysics, Inc.

Fig. 6–11 Cube

In any of the various forms of combination display, a system may be able to replace a surface of the display by the next section or slice "behind" it, then the next, etc. Or you can replace a surface with one "in front" of it, progressing through the data in the other direction. The system may be able to replace two, or all three, of the displayed surfaces at once. In any of this replacing, it can skip one or more each time, to move through the mass of data more rapidly.

Such changeable combination displays allow you to work your way through the whole solid mass of data in sequence, but always with the other surfaces to look at, whether you are changing just one, two, or all of them. There is a price to pay for being able to see this data in sequence. More computer operations are required to call it up than for one section at a time, so it may be slower. And only some of the systems are able to make such complex displays and run them in sequence. Those systems are usually the ones with more computing power and disk memory.

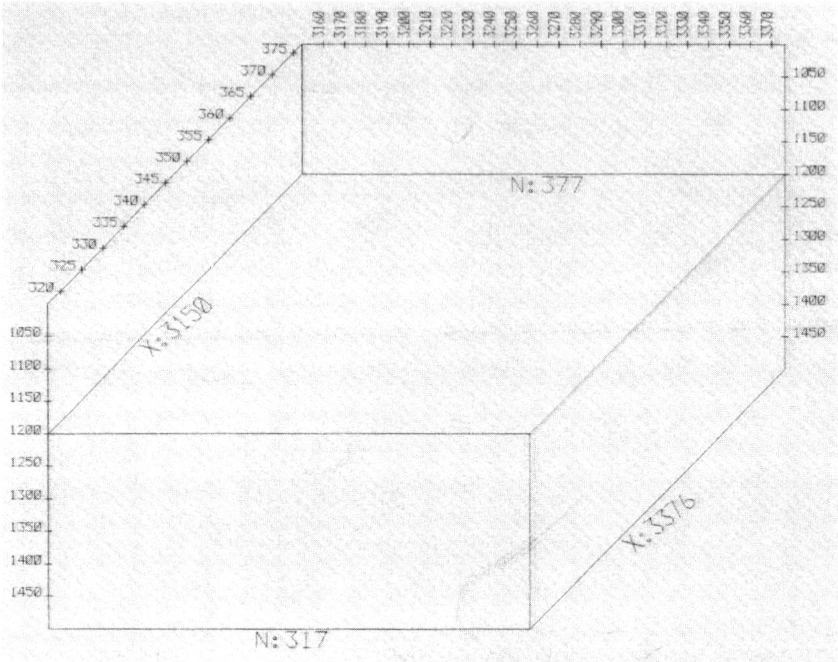

a Plan

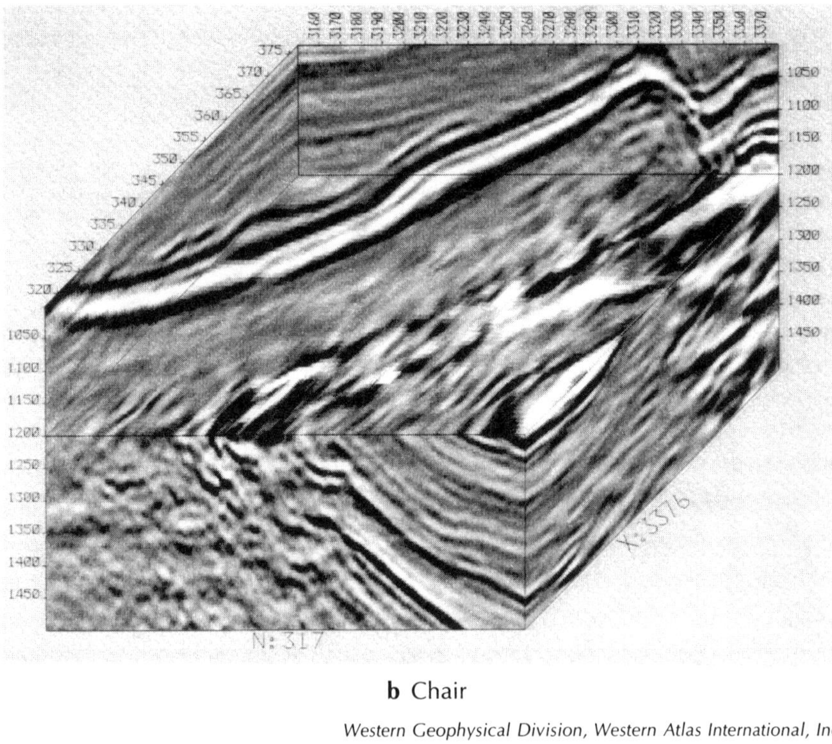

b Chair

Western Geophysical Division, Western Atlas International, Inc.

Fig. 6–12 Chair display

MOVIES

As I mentioned in Chapter 4 in discussing a first look at the data, a sort of movie can be shown by having the system display a section or time slice, then replace it automatically with the next one, and so on. It takes a little time for the computer to display a frame, so the "movie" doesn't fool your eye with the appearance of continuous motion, but it can be quite informative. The speed of formation of each frame depends on the computer power you have available. The system shows the sections or slices one after another as fast as it can display them. You see one take shape from left to right or from top to bottom, then it vanishes, and another one forms. With some systems these "movies" can be displayed automatically from the normal storage of data, but on others, for a movie to be shown rapidly and automatically, it must be created and saved as such on disk, and then shown from that copy.

When you are just starting to interpret a 3-D area, you can select a movie display of the time slices, and watch the screen show slices deeper and deeper into the earth. It's a heady feeling, seeing all this subsurface before your eyes. The movie is so informative that it is worthwhile to view it a couple more times, one of them backward so you feel that you are moving up through the earth. Then you may want to view other movies, looking through the area as a series of inlines and then as a series of crosslines.

An alternative is to show the data as a real movie, that is, with each image photographed on film or recorded on videotape, and shown on a regular movie projector or videotape player. Then the feeling that you are falling into the earth is enhanced. You have a feeling something like the scuba-diver's "fascination of the deep" making you reluctant to tear your eyes from the view that unfolds before you. But you would not be at your workstation, able to stop and go into another function whenever you wanted to examine something more closely.

This looking at movies has its main use in giving you an overview of the area, so you get an idea of the structure and the faults before starting to actually pick reflections. Faults in particular may show up best in movies, with subtle breaks evident to your eye only when you see them on a number of frames in quick succession. A movie can also help later, when you encounter some problem in interpreting, that makes you need to back off and look at the whole area again.

3-D PICKING

Picking a line is mostly done the same way in 3-D as in 2-D. A section looks the same in both kinds of shooting. But in 3-D there is so much data that you need some shortcuts in order to handle it all.

You don't necessarily have to pick all the inlines—the close spacing is shot largely so xlines and time slices can be made. The detail you need to pick depends on your purpose. After mapping at some interval, like every tenth inline, you will be able to see how well you have mapped for the kind of prospect you are looking for. If you have the feeling that you need more detail to see the picture, there are more lines you can pick. It's great to have the closely spaced shooting already done so you don't have to request more shooting. Maybe you will decide to pick more closely all over the area, or maybe just where you have found prospects.

After you pick a horizon on a loose grid, your system may be able to pick all the in-between lines automatically. The reflection may need to be pretty good for you to pick it reliably on those relatively far-apart lines, and for the system to pick the rest of the lines. Whether it can do it alone, or you have to re-start it frequently, will depend on the system you are using, and on the quality of the reflection.

A less automatic way of filling in after picking a loose grid of lines is to have the system contour the horizon on the map screen, and then have it mark the points where contours meet the sections onto the sections that you had not picked. Then you can use those marks on the sections to guide you in picking (Fig. 6–13).

In 3-D, there is a great temptation to speed up your interpreting with automatic picking. But since that works only where the reflections are continuous and good, some programs for automatic picking are kept conservative rather than daring, so they stop at the first difficulty. That way you can pretty well trust the program to not make foolish picks. You can allow it to go on as far as it can, and then fill in the blank areas on the map with your own picking.

Your system may have a function that lets you pick a reflection on one section, using any of the picking modes, and then have the computer display a copy of those picks on a nearby parallel section. You can use that as a guide in picking the new section. You can shift the copy up or down to bring it to the time of the reflection on the new line. You can then modify it, or, if it's in close enough agreement with the reflection, give a command to snap, and it will adjust to the reflection. You may even have a function that operates automatically in this way, so you can pick one point and then have the computer pick that reflection over the rest of the area. The horizon could be useful for flattening to help you pick deeper, more difficult horizons.

Autopicking alone needs good reflections to keep it from going astray. To use it on poorer or discontinuous reflections, you can give it a few "seed points", points you have picked, scattered over the area, and let it start at all of them, working outward in all directions from each. If your seed points are consistent, it can pick the whole area (Fig. 6–14).

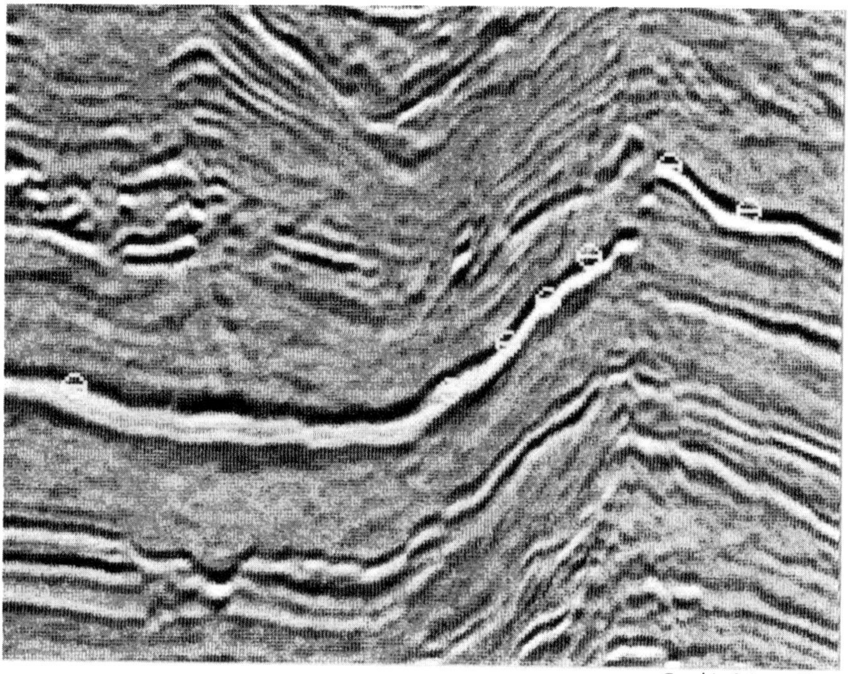

Graphic Science Corp.

Fig. 6–13 Map contours indicated on section

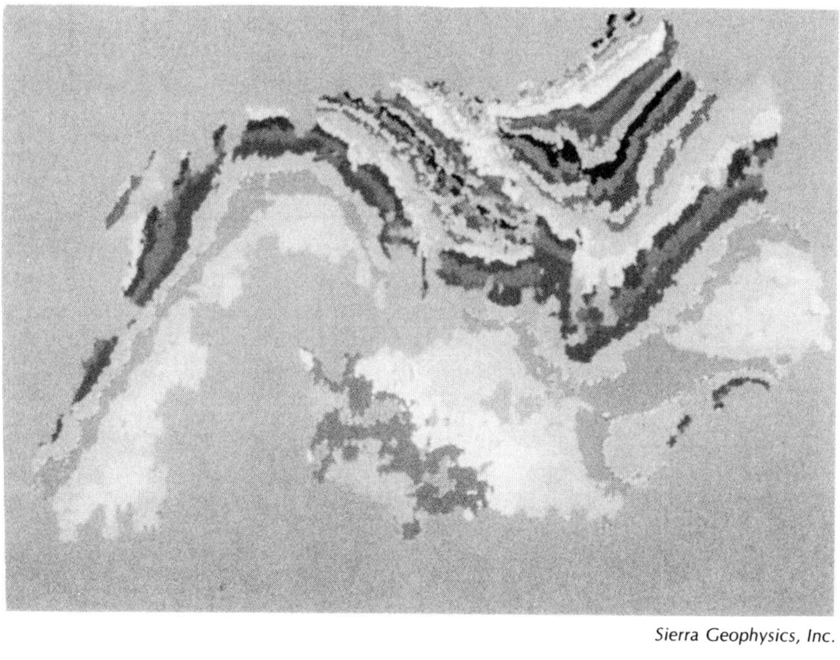

Sierra Geophysics, Inc.

Fig. 6–14 Automatic picking of area

The autopicking can start from a seed point, compare it with the traces next to it in four directions, pick those four traces, compare them with their neighbors, and go on, spreading its picking outward. With that extension in all directions, if it encounters a place where it can't pick, the continuing in other directions may cause it to pick all around the bad spot, leaving a hole in the picking. A map on screen with ribbons of color will show the gaps as blank (maybe black) places in the otherwise completely filled area.

CONSTRAINTS ON INTERPRETATION

Once, while shooting a large 2-D area in the Gulf of Mexico, we were having trouble interpreting. We decided the problem would be solved if we had closer control. So we assigned lines about twice as close together as we'd been shooting. After some of them were shot, I looked up my friend who was doing the immediate interpreting, and asked how the new close control was doing. With an expression on his face that told me life wasn't any easier, he said that it kept him honest. There were still problems, and he no longer had as much freedom to make choices between line intersections.

Interpretation of 3-D similarly "keeps you honest" by limiting your choices. It provides enough data to keep you from having to use your judgment so much in interpreting. It does this in several ways.

In 2-D, as long as you can make reflections tie at line intersections, you are free to vary your picks between those tie points. But in 3-D, an xline can be generated at every trace of the inlines. You probably won't generate all of those xlines, but if you did have all of them, there would be no places that could be picked between tie points. With fewer xlines generated, you aren't quite so constrained, but still have a lot of intersections. The effect of having all the xlines is evident on a time slice. The slice appears to have all the xlines in it, although it is made from just the inlines. By looking at it from different angles, you look down the inlines or the xlines. There aren't any points that don't connect to other points in both directions.

Another restriction is in the interpretation of dips of formations. Interpreting 2-D lines requires you to try to take sideswipe into account, somehow deciding how dip across the line affects the data. In 3-D shooting, migration too is three dimensional. Reflections are plotted in their correct positions (subject to the correctness of the velocity and other factors used in the migration).

Time slices put constraints on contouring. With data just on the scattered lines of 2-D, you have pretty much freedom to use your imagination in contouring. But a time slice has contour lines already drawn. That is, a reflection displayed on a slice shows where that reflection is at a certain time. That's what a time slice is, the data at one time. And that's what a seismic time contour is, the points where the reflection is at a certain time. The reflections on the time slice are usually so broad that they don't look like neat contours, but they show the directions the contours must take.

Your interpretation of faults is also restricted by the mass of data in 3-D. If you pick a fault on a section and then look at it on a time slice, you see its continuity in both vertical and horizontal directions. If a fault shows up clearly on both those types of display, the horizontal fault trace on the slice keeps you from connecting faults from line to line incorrectly, and the appearance of the fault on the vertical section keeps you from giving it the wrong throw. But if it is nearly parallel to nearby reflections on either section or slice, you won't see it on that display.

The constraints are all to the good, of course. When you are trying to learn about nature, the more information your data can provide, the less you have to guess, the better. Contouring from scattered lines may be a fine exercise of the imagination, but it's no substitute for knowing how the rock layers are shaped.

IS 3-D CLOSE ENOUGH?

The closer the control, the finer the detail you can have in your interpretation. Data from well logs is a scattering of isolated points. Contouring it gives a very general picture. A 2-D reconnaissance survey gives you, instead of separate points, a few lines to contour. That lets you make a much more detailed picture. Shooting more lines between some of those lines puts more detail in, so your contours lose their sweeping character and have more irregularities, more ins and outs. A 3-D survey carries the fineness of detail to what would seem to be an ultimate, with no blank spaces at all to be filled in by interpretation. But even that can be improved on. The rocks themselves have more variation than we get from a sampling of a Fresnel zone at every trace position. A dipmeter survey shows some dips that don't agree with seismic data. Why? Probably because the rocks have small irregularities that our relatively crude 3-D seismic investigation can't see. So we haven't reached an ultimate—we never do—we've just reached yet another stage of finer detail than we have been used to.

Another addition to the amount of information that we get in 3-D is already being applied in some areas—the use of three-component geophones to acquire compressional-wave and shear-wave data. The geophones record P, SV, and SH waves, yielding sets of data that can be compared to learn more about lithology.

COMBINING 2-D AND 3-D

In putting together a map based on seismic data, you will want to use all the data you have. The area may have some 2-D shooting in it and some 3-D. Should you combine the two, or just make one map based on the 2-D data and another using just 3-D? If the area you are interpreting is completely covered by the 3-D data, and a 2-D line or so is within it, the 2-D data would probably not add any useful information. You have the area thoroughly covered by the 3-D lines. But if you are interested in an area larger than the 3-D area, you need the information from any 2-D lines that extend past the 3-D coverage (Fig. 6–15). Combining the two is difficult in the same way that combining different 2-D lines is difficult. You have to compare, change polarity, bulk shift, change amplitudes—whatever makes the data match (Fig. 6–16). You would almost certainly make the change to the 2-D data, not the 3-D. The 3-D is probably the

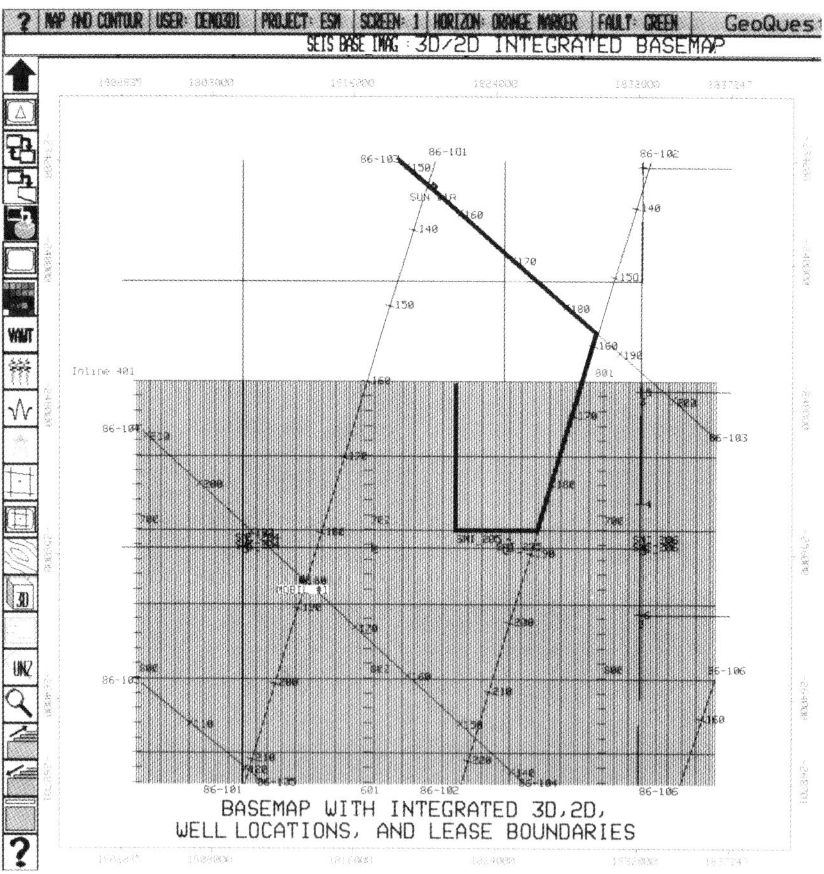

GeoQuest Systems, Inc.

Fig. 6–15 2-D and 3-D areas combined

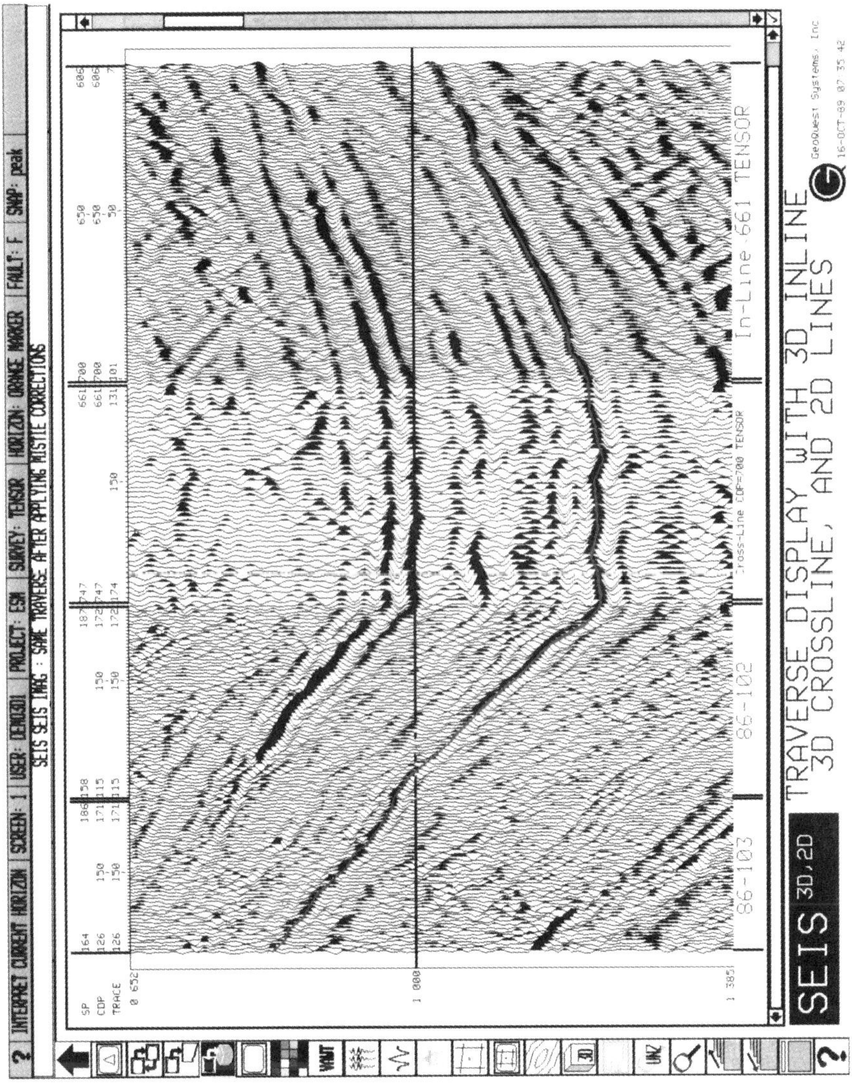

GeoQuest Systems, Inc.

Fig. 6–16 2-D and 3-D sections combined

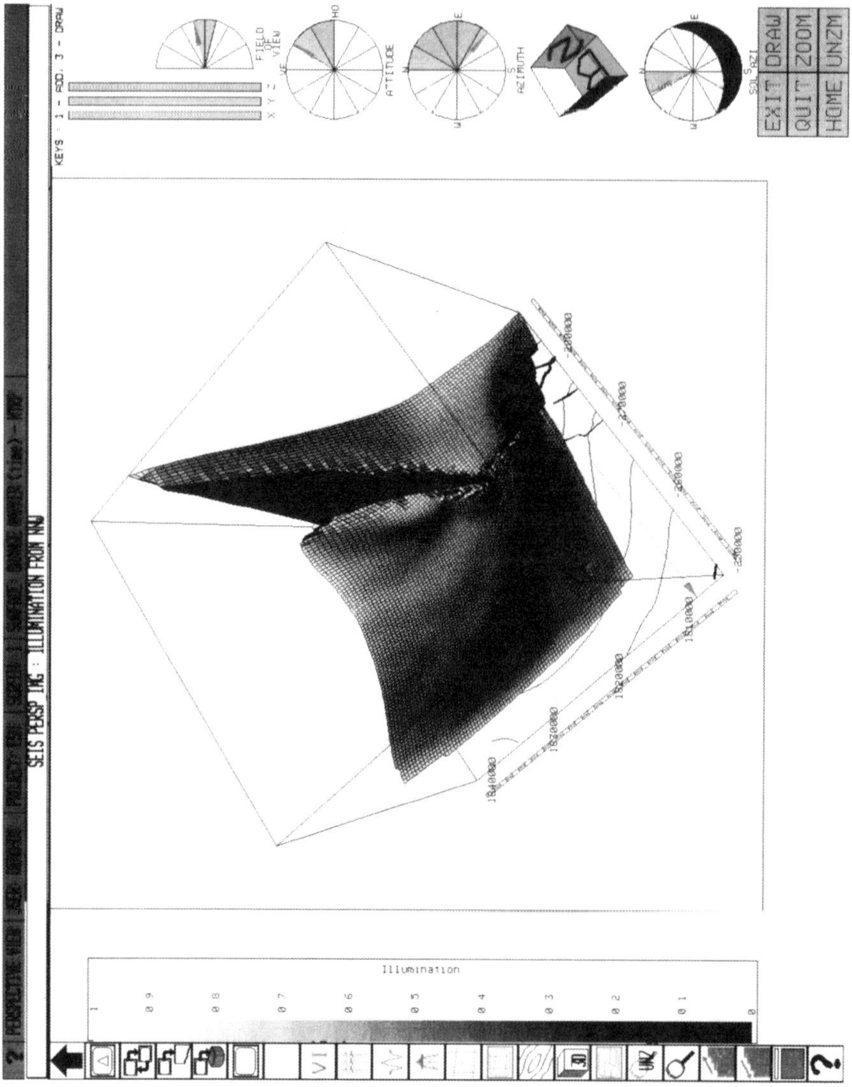
SEIS PERSP IMG : ILLUMINATION FROM NW
KEYS : 1 - ADD, 3 - DRAW
FIELD OF VIEW
ATTITUDE
AZIMUTH
EXIT
DRAW
QUIT
ZOOM
HOME
UNZM
Illumination

a

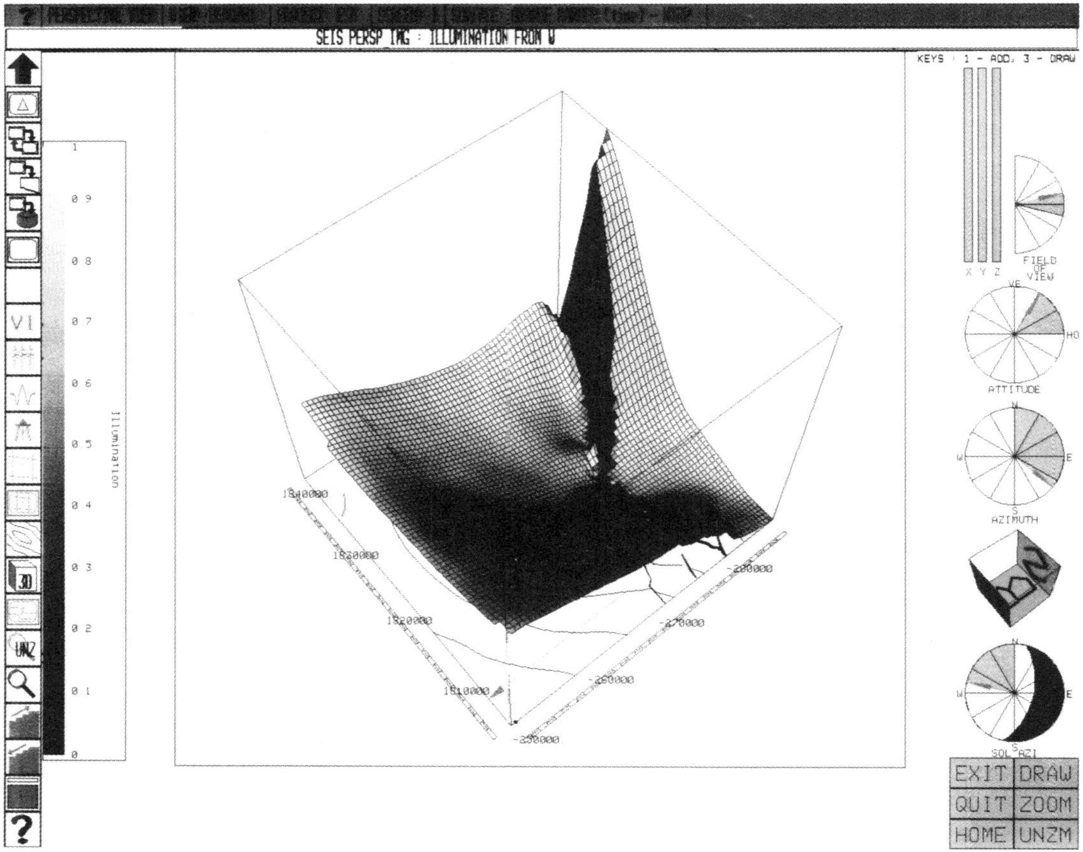

b

GeoQuest Systems, Inc.

Fig. 6–17a & b Wire-net display illuminated from two directions newest and therefore the best shooting in the area. Also, changing the 3-D would be changing a lot of points. It might be easy enough to change the uniform area, but you would probably prefer to make the change in the older 2-D area and leave the newer area as it is. Where a 2-D line runs into a 3-D area, you have plenty of intersections to match. If you have a problem matching the two, you can make an arbitrary line from the 3-D traces, a line that follows the same path as the 2-D line. Then you have two lines that should be duplicates, and can compare them anywhere along their length by displaying them together like two lines at an intersection. Once you've made them agree, you can pick from one to the other. Then you can pick and map the 3-D area and the part of the 2-D area that is outside the 3-D area.

Contouring such a combination area, either by computer or by hand, has the same problem that occurs in contouring well data. Wildcat wells are scattered widely, but in an oil field, wells are close together. Contours of the two arrangements on one map can look strange, with long sweeping contours where the wells are far apart, and with detailed contours wiggling around in the field. Spacing between contours changes too, from far apart in the non-producing part of the area to close together in the field. There isn't a satisfactory way to resolve those differences. If you decide to keep the regional look by just using one or two of the field wells, the map looks O.K., but you have thrown away some information. It's probably best to just let the map show the sharp change in detail. In combined 2-D and 3-D seismic contours, you also probably will want to let the difference show, with tighter contours in the 3-D area than outside it.

DIFFERENT VIEWS

Systems that allow picking on cubes and similar displays make it easy to see how your picks are shaping up. These displays are also often used to show features so non-interpreters can understand them. They are good for displaying in a meeting or including in a report. Making points clear to people who either aren't familiar with geophysics, or just don't know your area, is vitally important to obtaining help from people in other specialties, and to getting your interpretation accepted. If you find a good location for a well, and then fail to convince anyone that it is good, much of the value of your interpretation is wasted. As the power of computers increases and new programs come along, these displays could someday be the main form of data for interpreting.

Some systems can show wire-net displays from different angles, so they appear to be three-dimensional objects that are rotated. You can see them as though viewed from any direction, and from the same elevation or from above or below. You may also be able to change the apparent

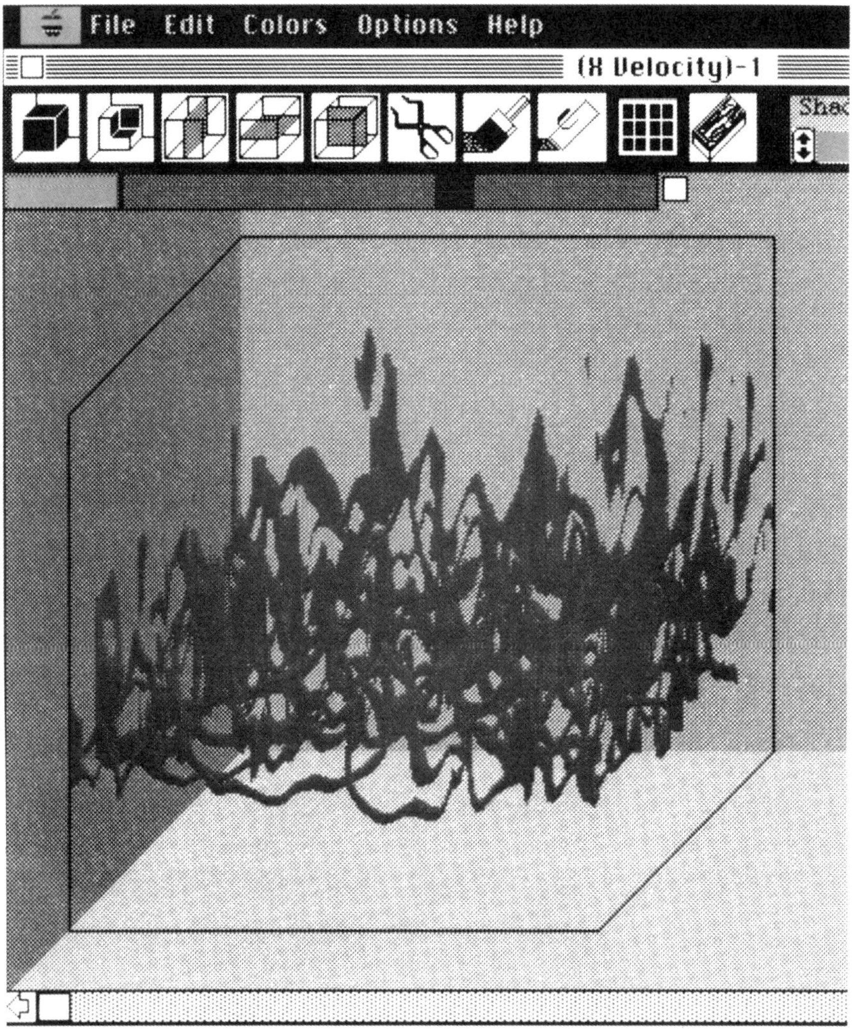

Visualogic, Inc.

Fig. 6–18 A different view
Copyright 1988, 1989 Visualogic, Inc.

direction of illumination, so shadows are cast in different directions, highlighting different features (Fig. 6–17).

On most systems you can limit a display of a section or slice to certain features by the way you assign colors. For instance, to see just the bright spots, you can assign the background color (probably white) to all but the highest amplitudes. A development that has already occurred, but as far as I know hasn't yet been applied to seismic interpretation, extends this capability to three dimensions. It displays a volume as a cube, then lets you remove parts, and displays the data behind them. That much isn't particularly new. Seismic interactive systems do that, at least to the extent of letting you remove a seismic section from a cube and display the next section. But this technique also lets you remove all the data that is displayed in a certain color, so you can see through it as though it was invisible. It is interesting to use this technique to make all colors but one invisible (Fig. 6–18). In a seismic cube display, you might make all colors invisible except the highest amplitudes, so the bright spots for the whole volume would appear to float in air. Viewing the cube from different directions would allow you to see the three-dimensional patches of bright spots from different angles. You might also use it to show certain frequencies, or certain velocities from a synthetic sonic log section.

7

Solutions to Seismic Problems

All the advantages of interactive interpretation are solutions to problems. Much of this book is devoted to covering them. Seismic work, people like to say, is conducted to solve geological problems. Yes, any problems about the subsurface are geological. But on the way to solving problems about the earth, there are many seismic problems to be solved—problems about getting the subsurface information and interpreting it.

THOSE SAME OLD PROBLEMS

In any advancing technology, there are many problems. With time the severe problems of one stage of the development are solved, partly solved, bypassed, or accepted as inevitable. The next stage also has problems, but they are different ones.

Some of our oldest problems have been pretty well solved.

In the early days of seismic interpretation, early in the period when interpretation was done on individual records as they came from the field, a major problem was in matching the time scales of two records. Timing lines on the records indicated the times correctly, but the spacing between the timing lines couldn't be kept uniform. So a necessary skill in interpreting was sliding one record along another, matching times as you went, and correlating reflections while doing it. Recording at a uniform speed solved that one. Nobody worries about it any more.

The reflections on paper records were affected by topography, as modern field monitor records are now. The time from a geophone on a hill to a reflecting rock layer is greater than from a geophone in a valley to a reflector at the same depth. So reflections appeared to be deeper under hills and shallower under valleys. To pick reflections across a record, an interpreter had to observe the up-and-down pattern of a good reflection somewhere on the record, usually a shallow reflection, and then look for other reflections that had the same pattern, allowing for the lessened normal moveout of the deeper reflections. Corrected sections took care of the problem for the interpreter. Data processors correct for the hills and valleys, and remove normal moveout, so a flat reflection looks flat on the section.

Other problems have been largely solved, but are still with us. We still have to be alert to them.

Multiple reflections, in bouncing back and forth between rock layers instead of just reflecting once from one layer, look like ordinary primary reflections deeper in the section, with their increased dip looking like the steeper dip to be expected deep in the subsurface. Multiples have been greatly reduced by CDP shooting and processing, but are still a problem in some areas. You don't have to be constantly alert for them, but need to keep them in the back of your mind, to consider when something looks strange.

In 2-D there is still the problem of migration that is only along the seismic line, even though there may be dip in other directions. And in both 2-D and 3-D, there are the problems of incorrect migration caused by a less-than-optimum velocity and by the assumptions made in the migration programs.

LOCATION, THE PROBLEM WE DON'T TALK ABOUT

All the interpretation you do is based on the locations of the shot points. If they aren't correct on the base map, your picks are meaningless. But surveying, like any other activity, can be done well or poorly. How reliable the surveying is depends on the surveyor's skill, the type and condition of equipment used, the accuracy of existing benchmarks, and other factors. Other aspects of seismic projects are more under control. Processing can in part make up for poor field operations. If the processing is poor you can have it re-done. And of course you can correct things you have done wrong in your own interpretation. But in seismic work it is normal to more or less accept the results of surveying.

To a seismic interpreter, an incorrect location of a seismic line is a problem almost too horrible to think about. It's more comfortable to assume that the locations are correct, must be correct because they are so important. But like anything else, they can sometimes be wrong.

Fortunately, in modern offshore work, the problem has been largely solved by integrated surveying—using several methods of radio surveying simultaneously, and combining the results into a coherent determination of locations. A land-based method like Shoran might be used for precision, and checked for general correctness by a longer-range land-based one like Loran, and those two combined with satellite measurements. A flaw in one method can be found by the others. In shooting on land, a degree of integration is obtained by taking some satellite fixes at key locations, for instance at ends of lines. Using old shooting, of course, brings in errors that were made before people started using integrated surveying.

Data from the surveying is processed, often by a special surveying company, to make the most consistent map possible from the information available. Those survey specialists can use notes from optical surveying, recorded data from radio surveying by satellite, Shoran, and the like. They can use air or satellite photos and topographic maps. They can relate the results to landmarks.

But there is a kind of information about surveying that is available to an interpreter, that the surveyors and the mapping company do not have. The special kind of landmark that you know about is not on the surface. It is the relative depths of layers of rock. Those are big, prominent, fixed landmarks. Suppose you are interpreting a 2-D project made up of shooting done at several different times by different companies, a type of interpretation you may often do. Let's also say there is a good, easily recognized reflection on the sections, one of those strong reflections you often find where there is a radical change in rock type. Maybe it is the horizon you are mapping. Now, at a line intersection you find the reflection at one seismic time on one of the sections and at a much different time on the other. The discrepancy is comparable to the one a surveyor would have if the surveying indicated a large cliff on the surface, at a place where a glance at the area would show that there wasn't any such cliff.

If there had been just a small difference between sections, the problem could have been caused by a polarity difference or by a difference in datum or some other near-surface correction. But if the difference in times of the reflection on the two sections is large, too great for that sort of problem to account for, it may be that the locations of the lines don't agree. If there is a large fault nearby, moving one line across the fault may bring it into agreement with the other. Or if there is steep dip, then moving one line until the reflections fit may do it.

In interpreting on paper sections, the adjustment is made by just marking new positions for the intersections on one or both sections, and plotting the line in its new position on the map. With an interactive system, what you can do depends on the capabilities of the system. Some systems allow you to shift a line's location, some don't. If your system doesn't allow that, then you could get the change made in the location database, maybe even get the surveying people to readjust the locations, fitting your observations in with the information they have.

But interpretation is usually done under some pressure of time. You probably don't have time to wait while the surveyors work with the data and come up with a new map, and then someone loads that map information into your system. All of that is more useful for future interpretations. While you are working on the area, you have to come up with an answer. You can pick the section, make notes of some reflection times, and add them to the map at the positions where you believe they belong. Your system will probably have a provision for you to add isolated points. If not, you may be able to treat them like well data, and enter them as such. That may involve loading them into the system as wells.

If your system allows you to move a line, then you have a problem in moving it correctly. This isn't to be done lightly. You are making a big change in some of the basic data. However you move it, it will have to agree with other lines it crosses, both in seismic times and in surface elevation or water depth. And you will need to make prominent notices to let people know what you did—on maps, in reports, and in some kind of annotation that goes in the magnetic or optical storage of the interpretation. And after all that, it would be wise to go back to the surveying specialists too. You have landmarks they don't have, but they have information you don't have. The two should be fitted together.

In a 3-D project, this problem does not exist. Or rather, the way the shooting is done all together as one unit, such a problem is so unlikely as to be almost impossible. And if by some chance a big location error did find its way into a 3-D area, it would not show up as such on the sections. The data is binned in uniformly spaced cells, and any location error would already have been absorbed in the binning. The problem might appear only as a reduction in quality of data.

If you are combining two 3-D areas, there can be a surveying discrepancy between the two. You would handle it the same way you handle a failure to tie between lines in a 2-D area. The main difference is that you have a lot more data, and the data does two things: it indicates how to move an area, and it provides many points at which it has to match.

INTERACTIVE PROCESSING

Another seismic endeavor, not a problem, but a solution to problems, is data processing. It has traditionally been done by computer, and done in batch mode, but now interactive processing is coming into the picture too. Interactive processing and interactive interpretation are starting to blend.

Data processing was done at first by running a processing step on the data, printing a section, looking at it to decide if the process had been effective, then running another processing step and looking at it on a section. Then that making of so many sections was replaced by making "data dumps", printouts of sheets full of numbers, and examining them. Now processing is coming back around to having sections to look at, not as printouts but as displays on screen. The immediacy of a screen display and the power of an interactive system make it possible to try varying parameters on a sample of the data until the best choices are found, and then send those parameters to a more powerful computer to guide batch processing of the whole area. The whole area could be processed interactively if there was any reason to do it. It would be tedious and would be a waste of time, but it could be done.

In normal batch processing, a whole area is processed in the same way to make it uniform for interpreting. But in a local area, a different processing might make the data there look better or help solve some particular problem.

Interactive processing, like interactive interpretation, gets much of its value from its immediacy. After running into a problem, you can see if a change in some variable helps. When you think up something to try, you can try it then and see the results quickly, instead of having to wait, and then maybe having lost the thought by the time you get the results.

That ability is available in several ways. Some interactive interpretation systems are connected to the processing computer, so you can try different kinds of processing on a part of an area with it. Other systems have some processing steps built into the interpretation system. And there are desktop interactive systems just for processing (Fig. 7–1). With any of these, you gain ability to resolve problems in your data.

You can derive a wavelet from the data and see it at once; mute out near-surface diffractions (Fig. 7–2); remove normal moveout (Fig. 7–3); make brute stacks to get a preliminary look at the data before completely processing it; try different varieties of deconvolution and see their effects on the traces (Fig. 7–4); make NMO velocity analyses quickly on the screen; try different bandpass filters (Fig. 7–5). You can completely process the small sample of the data on the system. Or you can just make some individual tests, and then have the parameters that you decided on used in the batch processing. You save time in analyzing, try more different things, and obtain

Parallel Geosciences

Fig. 7–1 Processing workstation

processed data that is better than it would be without the interactive system.

Velocities can be obtained by trying different velocities to remove normal moveout. These RMS (Root Mean Square, for the mathematical process used to derive them) velocities are available from any seismic data. This is one of the old methods, but is becoming more accurate because fast computers make it possible to economically make these velocity determinations at many more points on a seismic line than in the past, and because modern data enhancement provides better data to use in determining velocity. Among other improvements in seismic data, wavelet processing makes the data better, so you can pick velocities better. The better velocities make stacking better, so final sections are better in general, and reduce multiples more.

To make such a velocity determination on an interactive processing system or on an interactive interpretation system that can do processing, you can display CDP gathers and edit them by deleting bad traces and muting out unwanted energy like ground roll. Then you can have the system

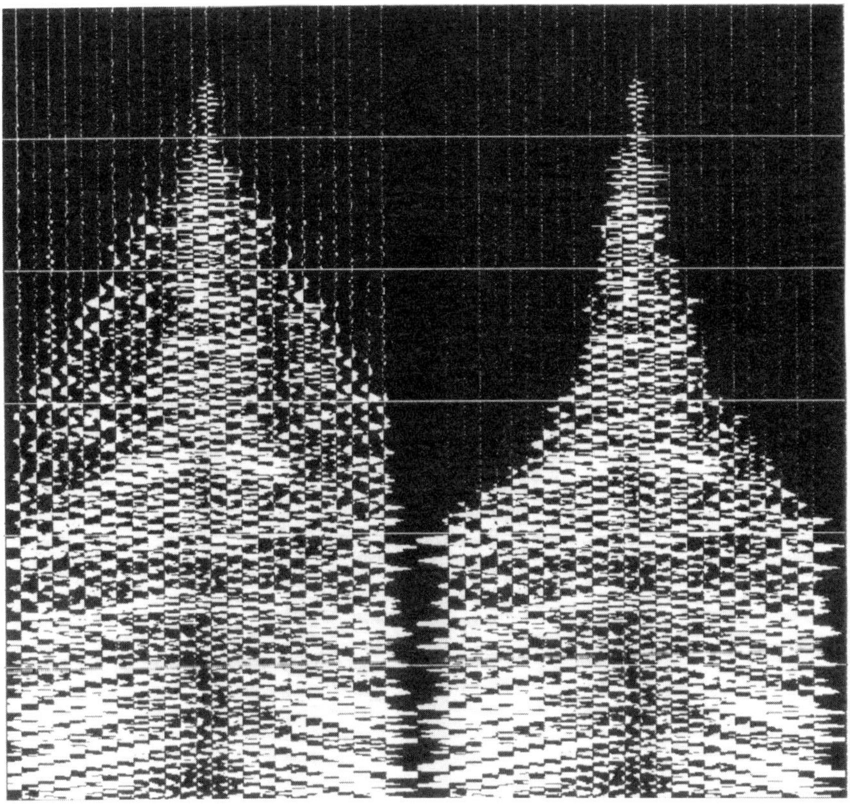

CogniSeis Development, Inc.

Fig. 7–2 Muting

display coherency of NMO-corrected reflections on a plot of velocity vs. time. This kind of plot usually has the coherencies contoured, so it is easy to see the differences from one velocity to another. Pick that interactively, by selecting points (Fig. 7–6). Multiples are easy to recognize. They are not on the main curve of primaries, but are straight below their primaries. A multiple has the same velocity as its primary, but is at a later time. You can draw velocity-time curves so quickly that you can have time to draw them at a number of points along a line or, for 3-D, in a vicinity. You can see which ones differ strongly from the others, and either reject those or accept them as indicating true lateral changes in velocity.

In Chapter 1, I wondered if, to be a good interpreter, you would also have to be an expert processor. The answer is probably that you would not, that you would have the ability to try variations on the processing,

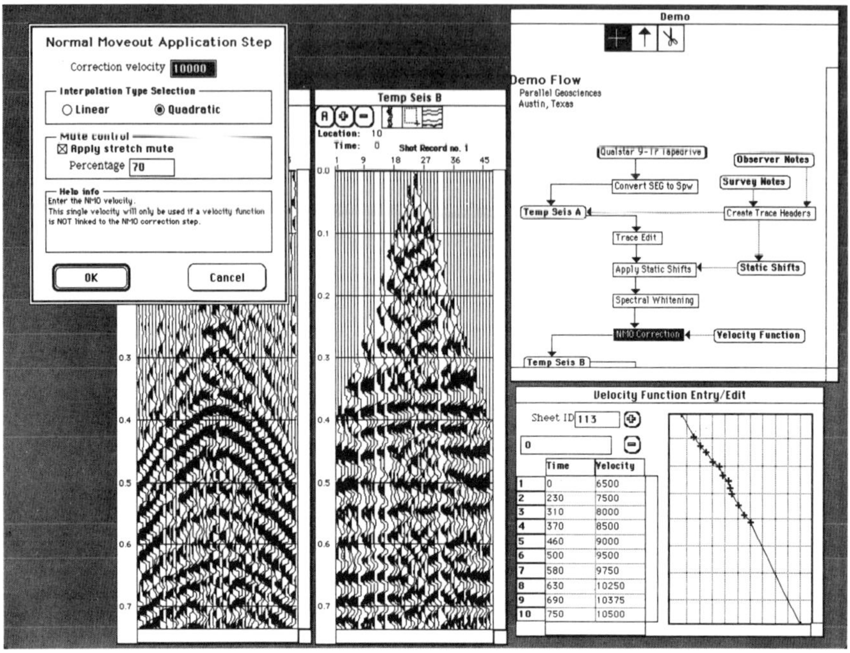

Parallel Geosciences

Fig. 7–3 NMO removal

and, if they were radical changes, that you would get the advice of a processor. Like any other specialty outside your own, it makes sense to check with an expert before plunging ahead.

FORMATIONS AND THEIR REFLECTIONS

With an interactive system you can investigate the nature of your data in ways that once were possible only by way of a processing center, at that time the only place in which seismic data could be handled by a computer. For instance, now an interpretation system can quickly obtain a wavelet from seismic data, the wavelet that the data appears to be made up of. The wavelet is useful in making a synthetic seismogram that matches the data. In some systems, the analytical functions are in programs that can be added separately. Some of the small systems specialize in this kind of investigation.

VELOCITY

Velocities determined from seismic data have in the past been unreliable and difficult to obtain. New techniques have made them more reliable, and interactive methods have made them quicker and easier to obtain.

The most conventional use of velocity is to obtain depths to seismic horizons. These velocities can be used to convert horizons, and whole seismic sections, to depth (Fig. 7–7).

Velocity can be displayed as an attribute, with colors for the different velocities, maybe overlaid on a wiggle-trace amplitude section.

In a more sophisticated use of velocity, velocities are determined at each reflection, using reflection strength to determine velocities relative to some known or assumed starting velocity. Then, in a reversal of the process of making a synthetic seismogram from a sonic log, synthetic sonic logs are made from seismic traces. A section made of these synthetic sonic logs is precise enough to show local variations of interval velocity that may indicate porosity.

IDENTIFICATION

Nowadays, a large proportion of seismic work is in development of known fields. In that, the identification of reflections with the formations they come from is usually not as much of a problem as it is in exploration, particularly in frontier areas. Reflections are usually fairly well identified before a field is discovered, and the identifications become reliably established shortly after the discovery.

Seismic data is measured in time, but well data is measured in depth. When you look at a reflection on a seismic section, there is no direct clue to its depth or to what formation it is reflected from. There are several ways to identify reflections with formations. If there are no wells in the area, you may be able to identify horizons using geological features, familiarity with seismic horizons in the general area, or actual seismic sections from a nearby area.

Sometimes you can find a geological feature of the subsurface, maybe known from outcrops, that is recognizable from the kind of reflection it produces. A thick carbonate layer immediately below discontinuous sands and shales could be such a feature. Or you may know the area so well from working other areas nearby that you can recognize some of the reflections. A few reflections, like the Wabamun of northern Alberta and the Batu Raja of Indonesia, are easily recognized, whether on screen or on paper.

Better, you may have a section from a nearby area that already has reflections identified, either a conventional paper section or a printout from a workstation. You could load that line in with your area to correlate with other lines on screen, but going through the whole loading operation may

7-4

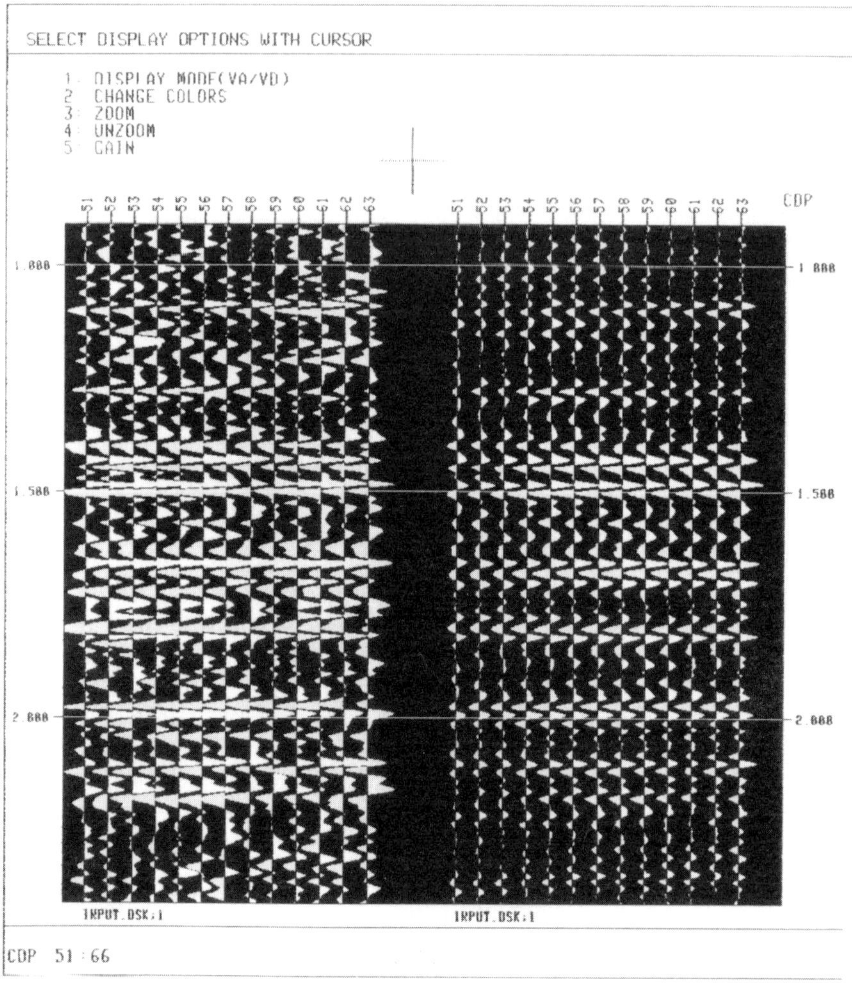

CogniSeis Development, Inc.

Fig. 7–4 Deconvolution

take longer than is worthwhile for just one line. From both areas you can get hardcopy of the screen images or playbacks from the processors, and match paper with paper. Or you can even hold a paper section against the screen and correlate.

If there are wells in or near the area, identifications can be made from information derived from the wells—velocities, synthetic seismograms, or VSPs (vertical seismic profiles). You can use any velocity information

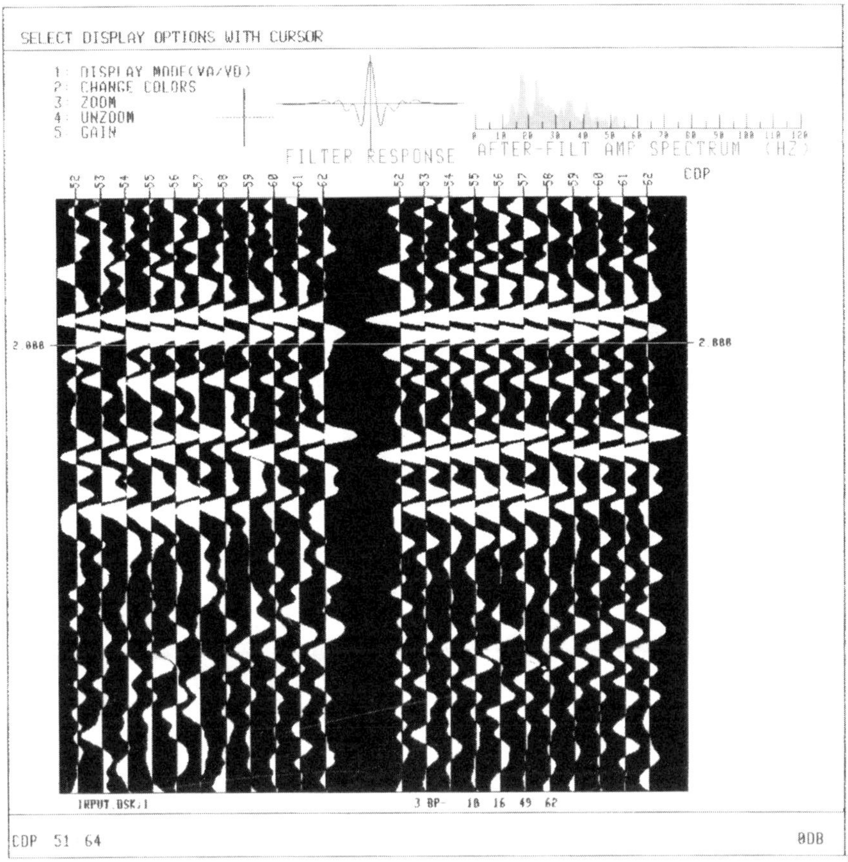

CogniSeis Development, Inc.

Fig. 7–5 Filtering

you have to determine depths to the reflections, and then match the reflections with formations near those depths in the wells.

If you have a well in which a sonic log, a density log, or both, were taken, and those logs are loaded in the system, you can use the system to make a synthetic seismogram (Fig. 7–8). A large part of the difficulty in comparing synthetics to sections is taken care of in the loading, by giving the logs the same datum plane and elevation-correction velocity as the sections have. After you give the command to make a synthetic seismogram, a number of choices appear. You can use any of several ready-made wavelets, or a wavelet you have the system derive from the seismic data. You can have the synthetic trace displayed in several forms side by side—maybe

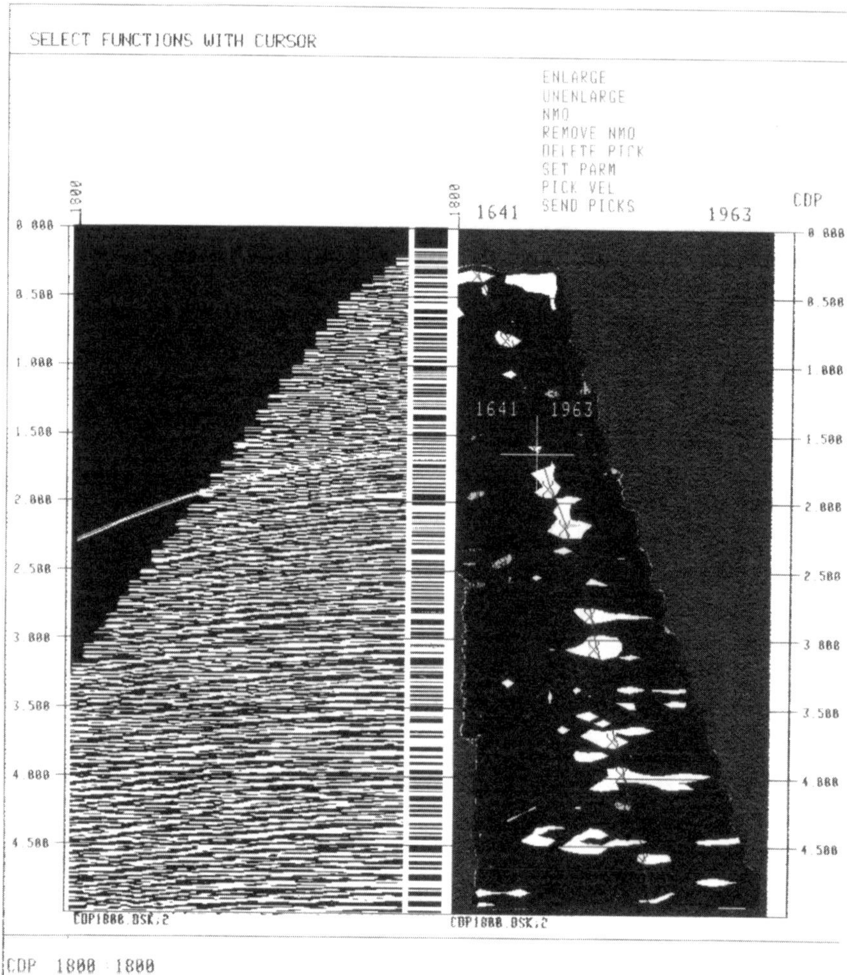

CogniSeis Development, Inc.

Fig. 7–6 Velocity analysis

in normal and reverse polarity, in various amplitudes, with and without multiples (Fig. 7–9).

You can insert the synthetic in a section (Fig. 7–10). Then it may be easy to correlate with the reflections on the section. Sometimes a synthetic can look disappointingly different from field data, but you can experiment with different wavelets and bandpass filters to make it look more like the section. Using a wavelet derived from the section is a great help in getting

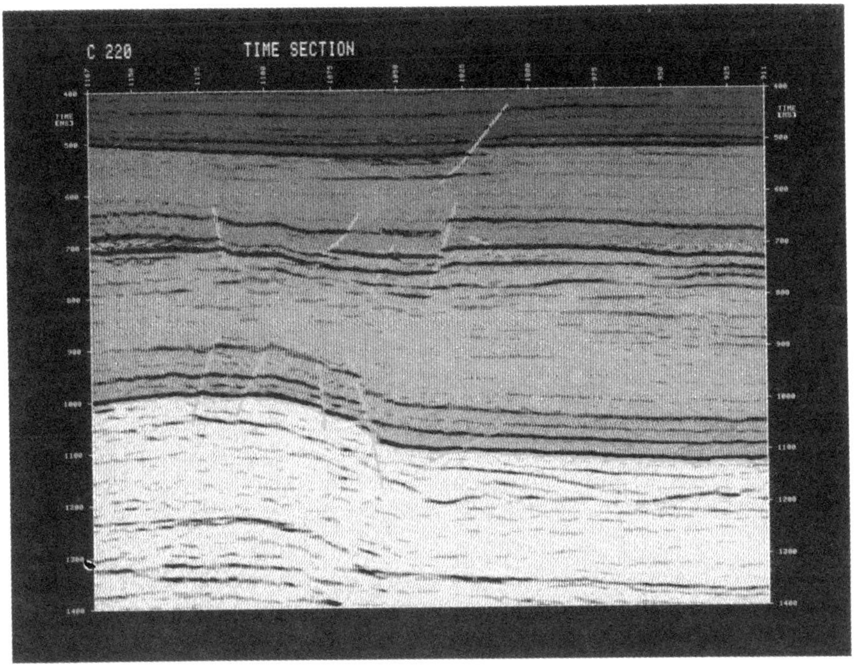

a Time section

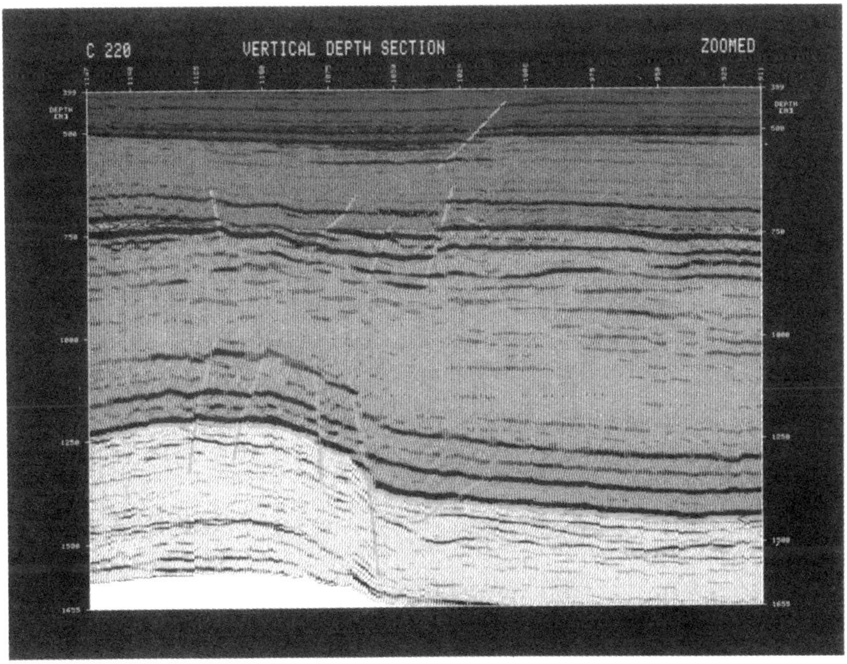

b Depth section

PRAKLA-SEISMOS AG

Fig. 7–7 Conversion of section to depth

wellsyn2	
PURPOSE:	Compute a synthetic seismogram from digitized sonic and bulk density logs, with optional checkshot corrections.
INPUT:	Input is a sonic log as two KINGDOM II signals with version numbers of 1 and 2 for depths in feet or meters and sonic times are in microseconds/feet or /meter accordingly. The SMT LOGPAK Software digitizes logs with a Houston Instruments HIPAD digitizer. Log depths are assumed to be measured with respect to the kellybushing (KB). The user may design a basic wavelet here or use one that has been previously defined and is in the scratchpad. Optionally, a bulk density log as two KINGDOM II signals may also be specified. Optionally, checkshot corrections may be applied to the synthetic. Checkshot depths are assumed to be measured with respect to the datum, where the synthetic two-way time is zero.

SELECT: PgDn ESC

OUTPUT: The synthetic seismogram data are stored as individual KINGDOM II signals. The user specifies the file name and wellsyn2 assigns version numbers 1–7. If check shot corrections have been made signals with version numbers 8 and 9 are also created.

Name	Item	
name.1	Velocity	These signals are
name.2	Bulk Density	input to other
name.3	Acoustic Impedance	super commands.
name.4	Reflection Coefficients	
name.5	Basic Wavelet	
name.6	Synthetic Seismogram	
name.7	Time-depth curve	
name.8	Corrected time-depth curve	
name.9	Drift (seismic time—integrated sonic time)	

DISPLAY: The output data are displayed as a seismic section. Press the F7 key and select to display the title. The title describes each signal in the display. Press F10 to continue.

SELECT: PgDn Esc

Seismic Micro-Technology, Inc.

Fig. 7–8 Menus

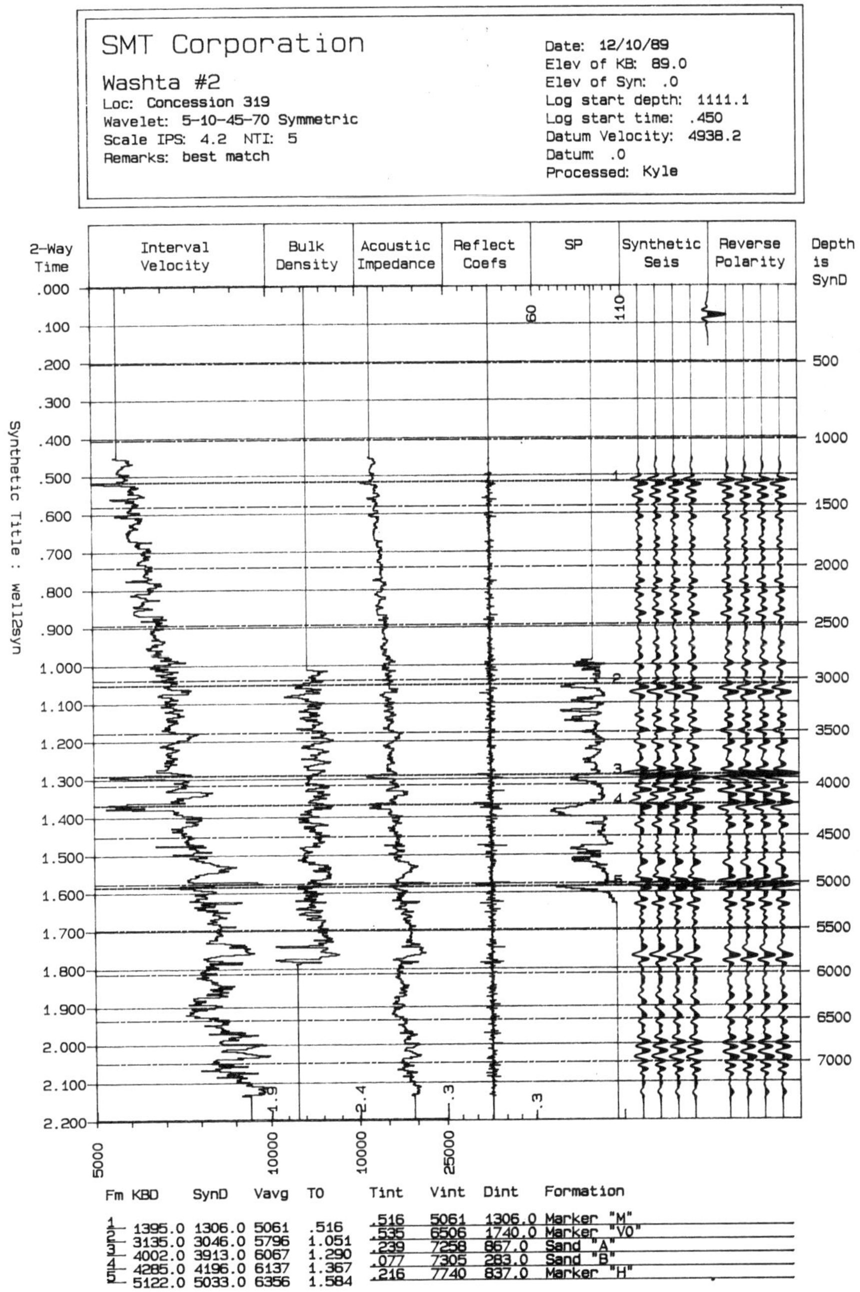

Seismic Micro-Technology, Inc.

Fig. 7–9 Synthetic seismogram

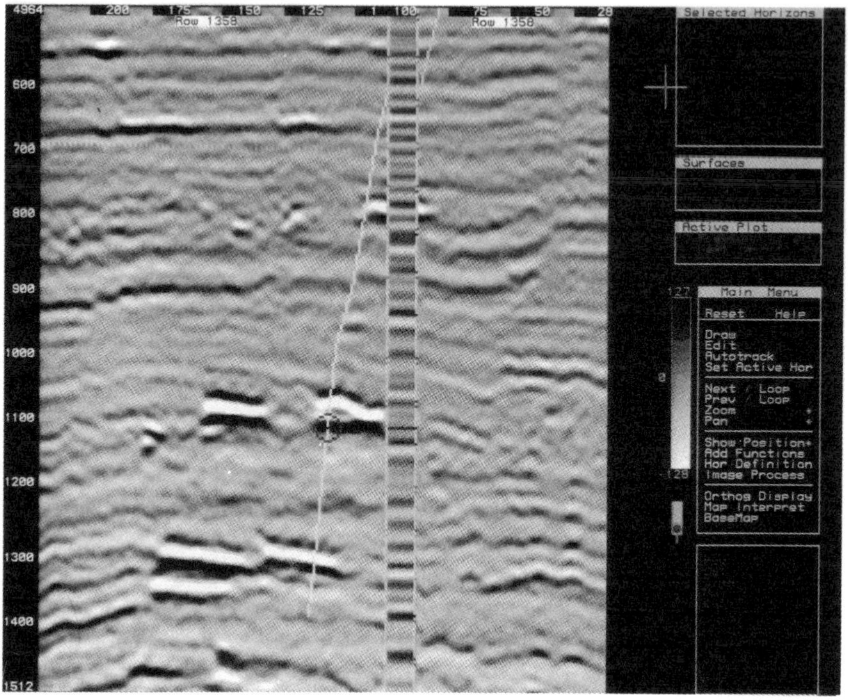

Geco Geophysical Company, Inc.

Fig. 7–10 Synthetic seismogram in section (with deviated well)

the synthetic to match the reflections on the section. In most cases you can make the synthetic fit the section well. Be a bit careful with this power to make the two fit, though. You can make the synthetic fit a nearby section, but you might also be able to make it fit one from an area with radically different geology.

A synthetic seismogram can be made more correct overall by adjusting it to fit a velocity survey. A velocity survey is made by using a seismic source near the surface while a geophone in a well is moved up from one position to another. It doesn't have as much detail as a sonic log does, but it is more correct overall. So a synthetic seismogram made by adjusting the sonic and density logs to fit a velocity survey is better than one without the velocity survey.

A VSP is like a synthetic seismogram made from a velocity survey, but with much more information. It too is made with a geophone at various depths in a well, but instead of having a sound source at just one place on the surface, it has sound sources at a number of different distances from the well. The resulting recording can be put in the form of a short seismic section, but with a known relationship to the formations. That bit of section can be displayed on screen adjacent to a regular section for correlation.

MODELING

Modeling is a general term for several functions. In an extremely general sense, any interpretation can be called modeling, but that dilutes the word beyond usefulness. The functions that are usually called modeling are what is technically called forward modeling. They work from the subsurface to a simulation of seismic data. The subsurface starting point, though, isn't known. That's why we do seismic interpretation, to try to find out what it is. So the modeling starts with assumed subsurface information. The assumptions can be guided by the seismic data. The process is a matter of iteration: start with seismic data, interpret subsurface from it, use that subsurface to derive a seismic section, see how much like the original section it is, change the subsurface, try again.

This is a good technique for determining velocities. Interval velocities between the picked horizons are part of the subsurface information.

The geology that is used in modeling may be known from the formation tops in a well, or it may be assumed from general knowledge of the area, or it may be the result of seismic interpretation.

KNOWN GEOLOGY

A synthetic seismogram is made from a log of reflection coefficients that is made from a sonic log and/or a density log. This is known geology to the extent that the logs are correct, not affected by washed-out parts of the hole or other problems. Anyway, in seismic work we usually assume that data from wells represents reality, because it is made from direct measurements of depth rather than from reflection time that then has to be converted to depth with uncertain velocities.

ASSUMED GEOLOGY

Instead of using a synthetic seismogram just at the well location where it was made, though, it can be repeated to produce a whole section of synthetic traces. And instead of making it from the well logs as they are, assumptions can be made that some part of the well is different from the way it really is. You can change a log by thickening a formation (applying the velocity of the formation to more vertical distance). Doing that progressively, you can make a section of synthetic traces in which, for instance, you make the zone of high velocity indicating a carbonate thicker on some of the traces to simulate a reef. You know how many feet of reef thickness you put in the synthetic section, and can see its effect on seismic data. Similarly, you can vary velocities to simulate lithologic changes, can repeat parts of the section to see how an overthrust fault looks, etc.

Going further with assumptions, you can dispense with real logs altogether, and think up your own geology. The synthetic section looks like a section from that assumed geology.

INTERPRETED GEOLOGY

The geology you start with doesn't have to come from logs or from your imagination. It can come from your interpretation of a seismic section. Then the synthetic section you make is a check on your interpretation. If it doesn't look pretty much like the section you interpreted, there is something wrong with the interpretation (the picks and the interval velocities you assigned between the picks).

RAY TRACING

Using synthetics works if the rock layers are nearly flat, but of course they often aren't, and then using vertical synthetics doesn't work. The way of adapting your synthetics to non-flat geology is by normal-incidence ray tracing, often just called ray tracing (Fig. 7–11). Seismic energy is bent by horizons it passes through and it reflects from horizons perpendicularly. Ray tracing calculates the angles of bending and reflection. The degree of bending caused by refraction at an interface is determined by the velocities of the two rock layers at the interface. You don't know the velocities in advance, but will determine them by trial and error ray tracing.

Normal incidence ray tracing simulates a section made with the geophones at the same locations as the shots. That isn't a way shooting is done, but is like a stacked section, in which the various paths of energy have been combined into one path, with the shot and receiver at the same place. Ray tracing can also be done with non-normal incidence, along paths at wider angles, going to all the geophones (Fig. 7–12). That is the way the shooting occurs, so it is like data before stacking. Its use is in analyzing normal moveout and velocity.

To determine velocities for a seismic section by ray tracing, display the section unmigrated. You need to start with an unmigrated section, because in the process you are going to migrate. Pick many events on it, including diffractions. Migrate the picks using the shallowest velocity. For offshore areas you'd start with the velocity of water. Then do ray tracing for the next layer. Some of the paths will be obviously wrong. Delete them, or adjust them to fit the rest. Then have the system smooth the horizon. Having that one level done, proceed to the next horizon, selecting, ray tracing, editing, and smoothing.

One advantage of ray tracing is that it migrates just a few horizons, and so is much faster than migrating all the data in a section. It can be

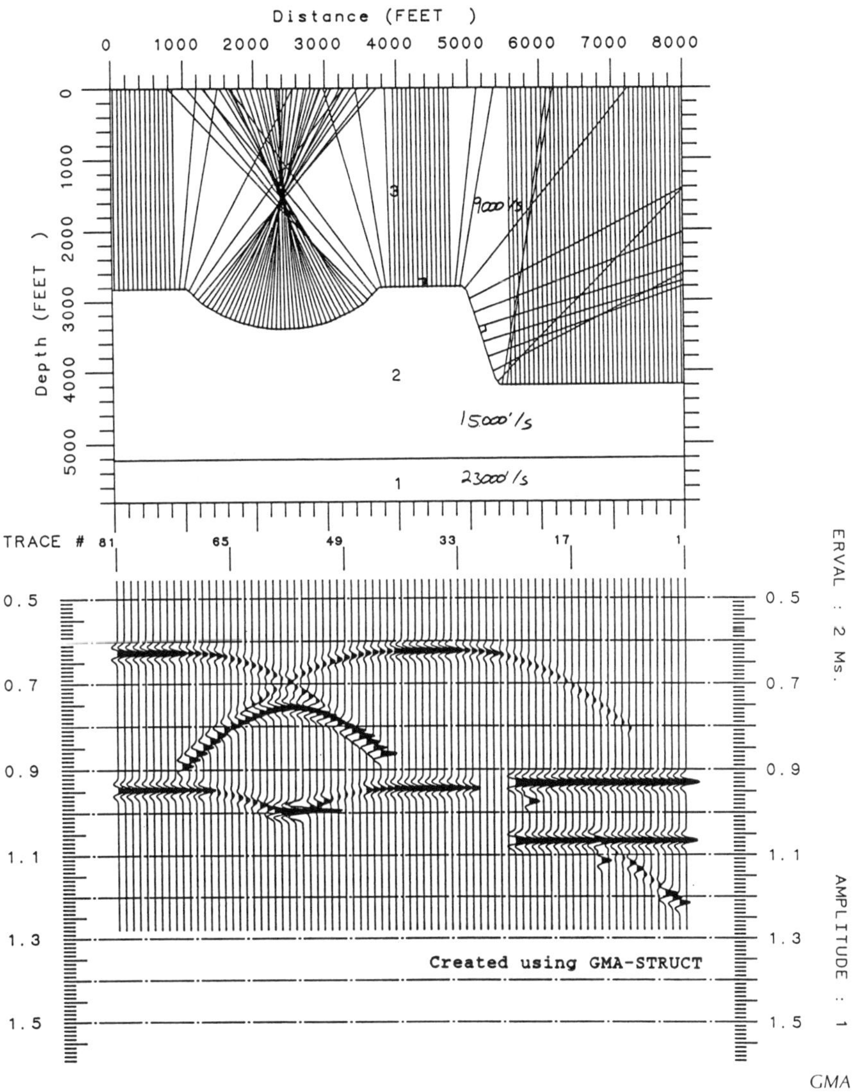

Fig. 7–11 Normal incidence ray tracing

used as a trial run, to see how a set of velocities works in migrating, before spending the time and money to migrate the traces.

USES OF MODELING

If you plan to shoot some subsurface feature, and already have some idea of its shape and dimensions, you can use that idea as the assumed

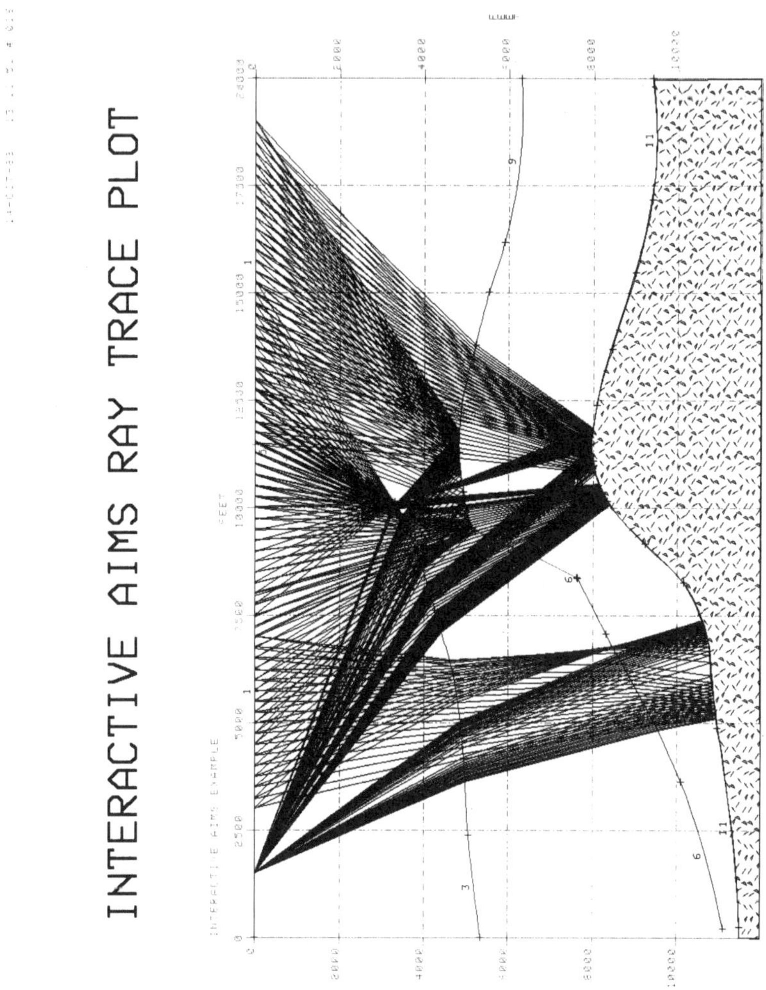

GeoQuest Systems, Inc.

Fig. 7–12 Wide-angle ray tracing

geology, and make model sections from it. You can make the model sections as though shot with different field techniques, to see what would best detect the assumed geology. You can thus experiment with factors like CDP fold and line length, to determine what CDP fold to use, and how far beyond the edges of a feature to extend the shooting, in order to map the feature effectively.

You can use modeling to decide the meaning of features you see on sections. In the past I sometimes encountered anomalous features on paper sections. The features were short extents of extra thickness in a reflection. At the time I had to guess that maybe they were sand bodies, maybe river channels. Now, with interactive modeling, I could put sand velocities into a model, varied to represent porosity and maybe gas content. From that I could obtain a considerably firmer idea of what I was seeing (Fig. 7–13).

That kind of modeling allows you to check an interpretation you have made and to try out alternate interpretations. It both makes you more sure, and lets you use your imagination and then see if your wild ideas could possibly be true.

Modeling to verify an interpretation is like tieing a loop. Having a loop tie doesn't prove you are correct, but having it mis-tie proves that you are incorrect. Having a model look like the original section doesn't prove the interpretation is correct, but having the model not look like the section proves that either the interpretation or the modeling is wrong.

INTERPOLATION OF TRACES

Interpolating traces is a process of inserting artificial traces between the traces that were obtained by shooting. The artificial traces are made by averaging nearby traces, or otherwise making them like the surrounding traces. The interpolated traces might be added in processing, or put in by your system. The purpose of the interpolated trace is to make the data more nearly continuous, more like rock layers, which are usually continuous. There are several uses for it in interactive interpretation.

Aliasing is a problem with any data that is obtained in discrete units rather than continuously. Seismic data is obtained in separate traces, which come from separate geophone groups and from separate lines. If features in the subsurface are smaller than the distance between subsurface positions of the traces, then the features can't all be detected. A small feature between traces will be missed, and if two small features are close together, they may look like one larger feature on the sections. They are aliased into one feature. The closer together the traces, the smaller the features that can be detected without aliasing. The same trace spacing that detects large features correctly may alias smaller features.

Aliasing also occurs vertically, affecting the features encountered in

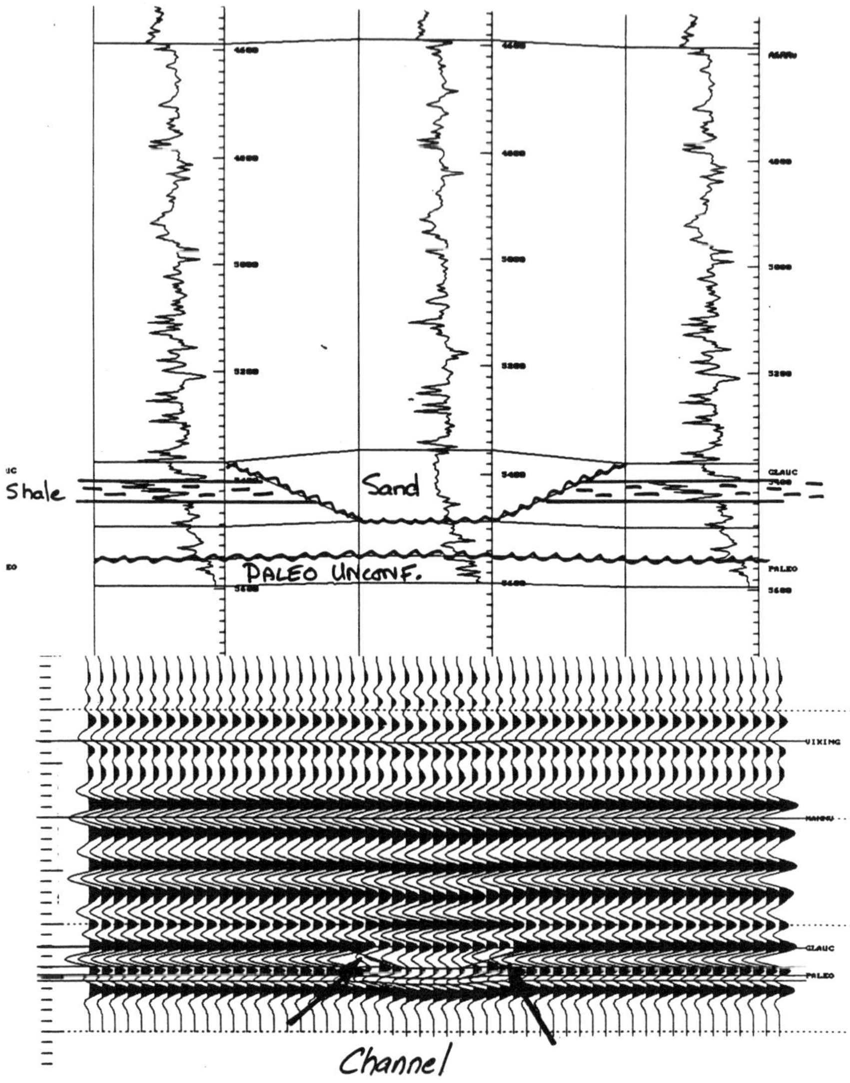

Fig. 7–13 Model of channel

one dimension by a single trace, with wide (low-frequency) events detected easily and narrow (high-frequency) events aliased. But the particular aliasing problem I am discussing here is horizontal, in which features small in area may be aliased.

A 3-D area may have been shot with the lines too far apart for accurate migration in the crossline direction. Adding new lines created by interpolation between the lines that were shot provides an effectively closer spacing of inlines, and therefore of traces on crosslines. The ability to interpolate traces can affect the economics of shooting a 3-D survey, by letting you shoot fewer lines and fill in later. The projects are so expensive that anything that can save some of the cost while still getting the needed information is well worth doing.

Field conditions can cause shooting to be so incomplete that some bins don't have enough traces in them. Interpolation can be used to add the needed traces, thus correcting for uneven spacing of geophone groups in the field. Ocean currents, islands, towns, and permit problems can cause the spacing of traces along a line to be variable. But plotting on a section is done with a uniform space between traces. So evenly spaced traces represent irregularly spaced groups. Measurements of distances and angles read from the section are not quite correct. And neither is migration performed as though the uniform spacing was correct. To get around that problem, traces can be interpolated at the uniform spacing, the original off-location traces thrown out, and the newly uniform section used for measurements and for migration.

For interactive interpretation of 3-D, your system may have a command that automatically sets the trace spacing of xlines to close to the same horizontal scale as inlines. If that function isn't available on your system, you will have to add interpolated traces to one type of line to make its distances about the same as those of the other kind of line. But making the scales exactly the same might require some non-whole number, like 1.3, traces added to one direction of line. That isn't possible, but you could try a "common denominator" plan, adding different numbers of traces to the two kinds of line. But then there would be so many traces in a line that it wouldn't fit well on the screen. Also, if the scales aren't equal, either because you do nothing or because your trace interpolation doesn't make the scales exactly equal, that adds distortion to any diagonal line you might construct, and to time slices and horizon slices. A circular feature would look oval on such a slice. The distortion probably won't make your interpretation poorer, but it isn't pleasing. And you will have to explain the distortion when you show displays to managers and other people who aren't familiar with the system.

A disadvantage of interpolation in interactive interpretation is that it increases the amount of data to be handled. The increase can be a problem in 3-D. If just one interpolated inline is made between each pair of lines that were shot, the number of traces to be put into memory is about doubled. And more than one line may be added between each two. The increase may sometimes cause the data to exceed the system's memory capacity. In that case the 3-D area might have to be broken into segments to be interpreted.

The search for smaller features has caused 2-D to be more and more like 3-D as the 2-D lines get closer together, especially when they are shot on a regular grid. Some 2-D detail lines have been shot as close as some 3-D. A 2-D area with fairly closely spaced lines (Fig. 7–14) can be given some of the advantages of a 3-D area by use of interpolation. Whether the data was shot in a uniform grid or not, traces can be added by interpolation on a uniform grid, as close as is used in a 3-D area. A grid of bins is formed (Fig. 7–15), and data interpolated to fill in the bins. The uniform interpolated traces are used, and the original traces not used. The binned data can then be given 3-D migration, with its greater correctness than

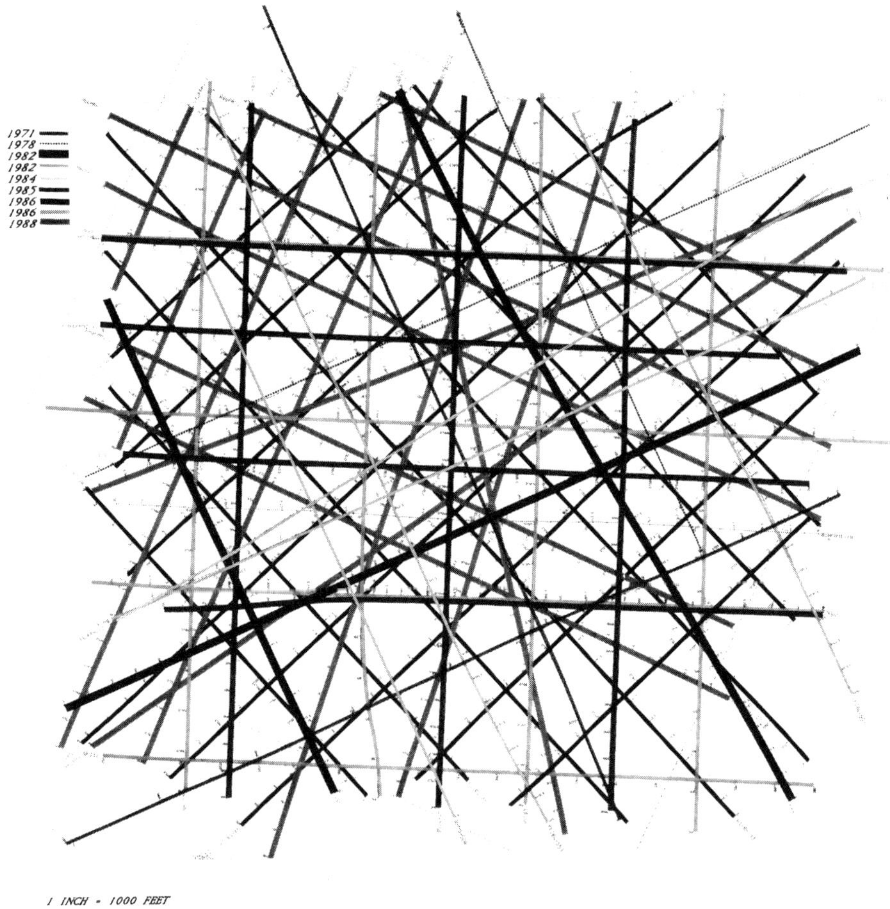

Teknica, Inc.

Fig. 7–14 2-D lines

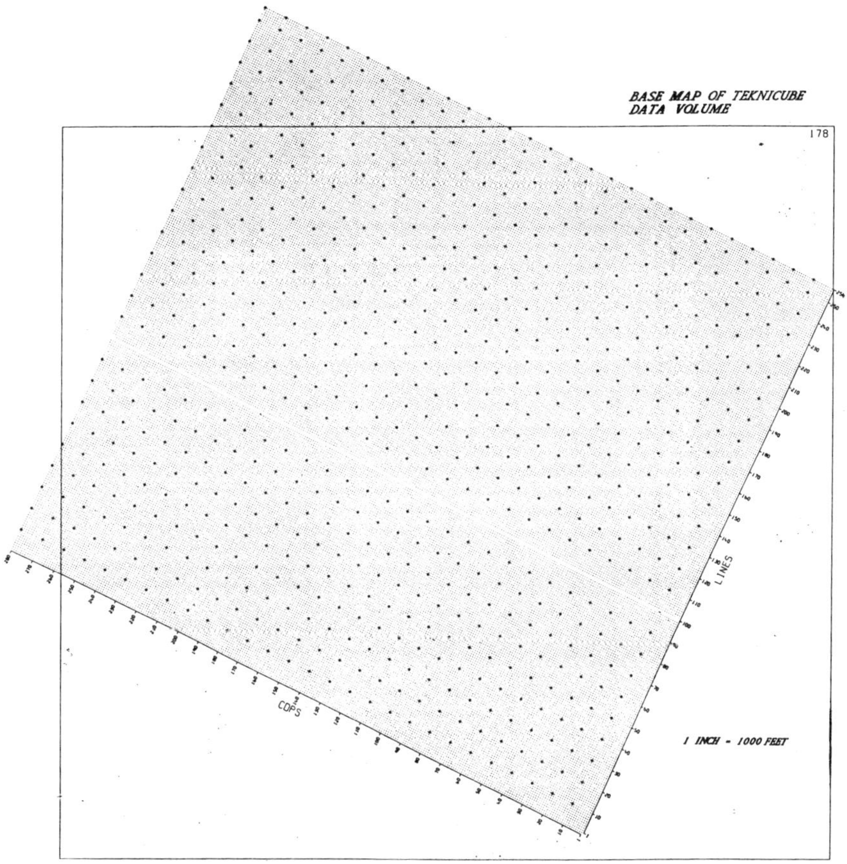

Teknica, Inc.

Fig. 7–15 3-D grid from 2-D

2-D migration could achieve. The same traces can also be used to form time slices. And lines can be made up in any direction. The 2-D area has become, in effect, a 3-D area, although not as good as if the data had been shot in 3-D. The interpolated traces are not real data, so no real information has been added. The interpolation of traces makes a reasonable filling in of the gaps. Those gaps are still places where no information was recorded, still places you don't really know about. But being able to use 3-D migration and to display time slices make the 2-D area yield better information than it would without the interpolation. You can do more with it, and find more oil with it.

INTERPRETATION BY PERSISTENCE

Some interpretation projects can be simple, easy, and quick. But most interpretations, even if they appear easy at the start, turn out to have some problem or other that makes them difficult. So interpreting becomes a matter of struggling with the major problem or problems of the area.

In interpreting either on paper or on a screen, the solution is arrived at in the same way—by working with the data so long and in so many ways that it all becomes familiar. In solving a complex problem, it may be necessary to have the data and the area so familiar that it becomes easy to remember where and under what circumstances a problem or feature appears.

Let's say you are working a large area interactively. You put a section on the screen and superimpose data from a well at the point where the line crosses the well. The match probably isn't perfect, but you manage to identify some reflections as being from certain rock layers. You may then display a line that connects that line with another well, and identify reflections at the second well. Then you pick those reflections from one well to another. At first one of the reflections mis-ties, so you re-examine the picking and the identifications at the wells. With some difficulty you get a satisfactory tie. This goes on for the several other wells there are in the area.

Then you start picking the reflections over the rest of the area. But there is a problem working from the east side of the area to the southwest. Automatic picking leads from the reflections you identified into some in the southwest corner that seem to be picked too high. You try jumping up in the corner and working back to the east. It seems to work on some lines, but not on others.

After a frustrating time, you change your approach entirely. You select a grid of a few lines and try working it like a new area. The same problem crops up, but on the lines that cross the change, you think you may have solved it. Then you start filling in the other lines in between, and your solution comes apart. You feel that you have wasted the time spent on the grid, and go back to the way you were working, but it breaks down again. You select one key line, and look at it in every different attribute display available. You pick across the bad spot. Displays of seismic attributes don't shed much light on it, but a plain change of amplitude setting seems to help. You decide to go over the grid again, but with the new display. And so on.

Eventually, you manage to resolve the problem and pick the whole area in a way that is consistent and also makes sense. You feel that you have a good new insight into the area, and that it fits in well with an idea one of the geologists had been expounding. Altogether, it is a satisfying interpretation.

You may think the problem was solved easily when you finally used the right approach, and think that if you had just used that one in the first place, you could have got the interpretation done much sooner. No, probably not. It wasn't the final approach that solved the problem. The solution came from using that approach **after** you had butted your head against the problem so long, and worked on so many lines so many times that you knew the area extremely well. You could remember where the data showed the problem in one way and which lines had the problem worst. It was a familiar neighborhood to you. You had become an expert in the area. No wonder you finally saw the answer.

So don't regret the time "wasted" in the efforts that didn't work earlier. You needed to spend that time learning the area.

THE TAKE-IT-EASY APPROACH

As you work on interpreting an area, there is great value in pausing occasionally to "browse"—to relax and look at the data for things other than the main purpose of the interpretation. This depends on the time available, of course. If you are on a tight schedule, for instance to get a structure map made in time for a meeting, then you need to stick to that goal. But in most interpretations you are looking for oil. In a particular area something is known about the conditions under which oil has been found, so you will be primarily looking for a repeat of those conditions. But browsing will help you notice other phenomena. In an area where the oil found so far has been in structures, you may notice what looks like a reef. Or you may happen on a group of flat spots that might help you decide what part of a feature to drill. Or the unexpected discovery you make may be negative, pointing out something wrong about the assumptions you had been using.

Browsing often works best when you are on your own time—when you drop by the office on a weekend, or hang around after quitting time, maybe while waiting on some evening activity. At these times, you are idle, wanting something interesting to look at, and your mind is free to generate ideas. You can start up the system and call up sections at random, or check on something that puzzled you. You may make attribute displays that you didn't want to take the time for during working hours. All the abilities of the system are available for idle speculation. You may learn some startling things.

THE TRY-ANYTHING APPROACH

I once read about a fine way of gaining information—trying just about any technique available on a set of data, whether intended for that kind of data or not, and seeing what happened.

The try-anything approach is the opposite of doing things according to theory. In seismic interpretation there are a lot of imponderables, a lot of things theory doesn't account for, especially a lot that isn't accounted for by the amount of theory you might happen to know if you aren't involved in research. Or the theory may not fit the data as well as was thought. Or the data may not be just what we think it is. Anyway, trying things doesn't deny the theory, it's just a matter of playing around with the data to see what you can find out about it. If something useful comes from it, the relation to theory can be investigated later.

What are some things that might be applied to data more or less at random, with some chance that they might be useful? Well, there are many things that you can do quickly and easily, so you don't waste much time if, as is most likely, they don't produce useful results.

Suppose you have an area with a few wells scattered in it. Some of the wells produce, but you haven't seen any difference on the seismic data between them and the ones that don't. Try a flattened amplitude display at the level of the pay zone, and a frequency display, and a phase display. On each display, try a variety of color arrangements. See if any of the displays show the producing wells in a different color from the dry holes. Never mind why for now, you are just looking for any relationship. Try the same thing for other horizons related to the one that produces, horizons just above or below it, or a farther one that appears geologically related. Make some time interval maps between horizons near the pay.

If you find a relationship, especially one that shows a clear-cut difference between good wells and dry ones, you may be onto something. If you aren't yet confident enough to recommend a well on that basis, point it out to other people you work with, describe it in a report, and wait till another well is drilled in the area. If it agrees with your observation, you have more ammunition—you can point to the report to show that you predicted correctly. If it turns out to work reliably enough to drill wells on with reasonable (not necessarily perfect) success ratios, your company can make money. And you still don't need to establish the theoretical basis for it. Plenty of thought will be applied to that task later.

I didn't list many procedures. Fine. You think up some. Try them.

ALTERNATE VERSIONS

One use of interactive interpretation is in producing alternate interpretations. Usually, in interpreting, you should decide on a version and stick with it. If there is guessing to be done between ways to interpret the area, you are the person best qualified to make that guess. If you leave the decision to others, you are in a way asking them to do part of the interpreting for you, when they aren't as well equipped to do it.

But there are times when it is useful to make two interpretations. The foremost of the times is while you are working on the area, to help you make up your mind as to which version is better. There are other occasions, maybe when some specific bit of information is uncertain. And sometimes it is worthwhile to get other specialists to help with the decision. A geologist's knowledge of other features nearby, for instance, may make the critical difference.

An interactive system can help you make more than one interpretation for the same reasons it can do some other things for you—because it is so fast, and can store so much information.

If you are picking a reflection, and come to a choice, a place where you aren't sure where to pick, it is easy to try both choices. Just assign a new horizon name to one of them, and go on picking. Depending on the system, you may or may not be able to combine that horizon with the first part of the horizon you started with. If the system will do that, then you can have two complete horizons based on the two choices. You can have them both contoured and compare them. If your system won't combine horizons, you can get around it in one way or another. You can call the picks three horizons: the first part and the two alternates of the rest of the horizon. Then on a map you can have the system overlay one of the alternates on the first part. They won't overlap because they cover different parts of the area, so the map will appear to be one map of the whole area. The ribbons of color will, if the data is 3-D, look like a complete contouring of the area.

If you need to have the system make a truly contoured map, either because you are working a 2-D area or for a report or a closer look at detail, the two won't be contoured as one unit, but separately. The best way around that problem is to have two complete sets of picks. Let one of the alternates be called the same horizon as the first part. Then, for the alternate that has a different horizon name, call up the sections and continue its horizon into the first part, exactly following the original picks. You may have to do that manually. Or your system may have a provision for moving a copy of a horizon to another level on the section and modifying it to fit the reflection at that level. If it does, just use a copy of the first picks, and don't move or modify them at all.

In 2-D areas, there are often different ways the same data can be contoured. That is a large part of the art of hand-contouring—choosing which of two equally correct ways of contouring a feature to use. In interactive interpretation, the contouring is automatic, but subject to editing. You can hand-contour part of the map, and have the system replace its contours with yours. Then you can have it smooth yours if necessary. Or you may be able to hand-contour and then have your system do its contouring, guided by the contours you made by hand. Or, in an open space between seismic lines, you can add a reflection time at some point, as though a well was there. Adding an arbitrary time can, of course, be dangerous. It amounts to influencing the data to suit yourself. But in this use, it is just a way of choosing how to contour, a way of communicating your contouring decision to the computer. The addition is O.K., it is just interpreting, but you need to be cautious. It could be harmful if you failed to document the addition with a note in a report. Later, someone could assume it was real, derived from some stray well, shot point, or something, and not realize that it could be changed. Anyway, with the ability to influence contours, you can turn out several contoured maps of a horizon, and then see which looks most reasonable.

There are also a number of choices you can make in the assumptions the program is to use in contouring. You can specify the contouring grid, the degree of smoothing, and other parameters. Changes in these factors can result in quite different contouring. They can sometimes give you useful alternates without your having to do anything to the data.

MOST TOLERANT POSITION

A useful interpretation concept is finding the "most tolerant position" in a prospect. That is, if several factors influence the prospect, finding a drilling location that is fairly good in most of those respects, and is not condemned in any, rather than selecting a location that is the best in one respect, and ignoring the other factors. Interactive systems, with their ability to produce several versions of an interpretation, allow this principle to be applied to the different versions.

If you have been able to produce maps based on different picking or contouring decisions, and they all seem to have some merit, a location that is reasonable on most or all of them, and doesn't look bad on any of them, may be the best place to drill. That method may be considerably more likely to find oil than deciding on little evidence which interpretation is best, and then basing a location on it regardless of what the others show.

The most-tolerant-position idea works particularly well with faults in 2-D. If there are a number of faults in a 2-D survey, and you aren't sure how to connect them (you are almost never sure), then recommending a location according to one interpretation of the fault pattern—often one guess—can easily lead to drilling a dry hole. Connecting faults from line to line is often the worst problem in 2-D interpretation. You can connect them one way, make perspective fault displays and contoured horizon maps, store that information, and then connect the faults in another way and do the same. Working on paper you'd never have the time for much of that, but the computer does the displaying and contouring quickly. You can keep trying until you get a clearly most reasonable pattern. Or you may want to recommend a well, but not be sure which fault pattern to base it on. You may of course have to use one preferred version as the primary one in reports to keep from confusing people. But if you can find a location that looks fairly good no matter how you connect the faults, then you are much more likely to find oil. Having made the different fault interpretations also prepares you to answer people who ask if you couldn't have drawn the faults in another way. You can say, "Yes, I tried it, and it looks like this."

RE-INTERPRETATION

The introduction of interactive interpretation is an advance like data processing was when it was introduced. Data processing made a great difference in seismic work. And since then developments in processing have been so rapid that, as a rule of thumb, you can say that any data processed as long as three years ago could be considerably improved by re-processing. Advances in interactive systems will continue to improve seismic interpretation, in a way analogous to the advances in processing.

During most of the time the "three-year rule" has applied to data processing, interpretation techniques have stayed the same. You might try to do a better interpretation than had been done before, but, unless there was new data to work with, like new wells or new shooting in the area, all you could do was try to be smarter, more careful, or more imaginative than the previous interpreter. Now, though, interactive interpretation is such an advance that it is worthwhile to use it on any area that has been interpreted only on paper. And interactive systems also are advancing rapidly, so there may come to be a three-year rule for interactive interpretation. Or, because systems change so fast, the rule for them may turn out to be a two-year or even a one-year rule. As systems improve they yield more and more clues about the subsurface.

If your company has a large accumulation of seismic projects that have been interpreted through the years on paper sections, the interpretations can be put into an interactive system by digitizing the picks. They should be thus quickly made available for calling up on interactive systems. But any of those areas of particular interest should also be re-interpreted as soon as they reasonably can be. It can be useful to determine which of those areas are attractive in other respects—land ownership, new pays discovered in their vicinities, changes in regulations, etc.—and then interpret those areas interactively. And any time you decide to re-process data, you can best take advantage of the improved processing by also using interactive interpretation.

This book contains many reasons for using interactive interpretation for a current project. The reasons apply just as well to old data. Also, although the old data is mostly 2-D, there are a few 3-D areas that were interpreted before interactive systems came along.

There are a number of improvements on the interpretation that you can obtain by using an interactive system.

You can look over the fault pattern again. Different displays of a section can make faults show up better. A wiggle-trace display with both peaks and troughs filled in has twice as many filled-in events to be broken by faulting, so the faults may show up more clearly than on the wiggle-VA of most paper sections. A display of phase may also make the breaks in reflections more evident. So may a change in the proportion of horizontal to vertical scale. And an interactive system may enable you to do better at relating a fault indication on one section to one on another. The faults may be picked more correctly on each section, and also connected more correctly from line to line.

If there are velocity problems in the area, an interactive system may allow you to analyze the velocities on a representative line or so with its modeling function. That information can be used to make more correct depth determinations than had been made previously. And the analysis may call for re-processing, to use the better velocities for stacking and migration of the data. Different attribute displays can show flat spots, bright spots, and frequency changes that weren't seen on the paper sections. If the data is of high enough quality (shot with modern enough methods and equipment) for the techniques of seismic stratigraphy to be applied to it, interactive methods will make that more effective.

Pity the poor managers. Faced with requests to spend money on new projects, they tend to ask, "Will this shooting project (or this processing) be the last one you'll need for this area?" In the enthusiasm of the moment, you may think it is, and may tell the boss that. Or experience may prevail, so you may point out that you can't know what new methods will come along. Now the question can also apply to the latest interactive interpretation and the equipment to perform it on. You can never guarantee that the project you are proposing will be the last one needed. People keep inventing devices and techniques that tell more than the old ones did. And if a company doesn't use them, it will fall behind its competitors that do. Those competitors will be the ones making most of the oil discoveries.

8

Solutions to Geological Problems

There are a number of geological problems that interactive systems (Fig. 8–1) help you solve.

THREE DIMENSIONS UNDERGROUND

A problem that seismic interpretation has had until recently is that the subsurface features are three dimensional and seismic data was not. Early seismic records were "jump correlation" type, isolated shot points. Each record was essentially one-dimensional, like a well log, and the interpreter had to correlate from one to another, just as geologists do now from well to well. Later seismic data was shot in lines, yielding more or less two-dimensional data. This was much better, but still left a large part of interpreting to consist of judgment in contouring between lines. A rock formation that was being mapped might slope up and then down between seismic lines. And, in an area in which formations had steep dip, the seismic

Sierra Geophysics, Inc.

Fig. 8–1 Workstation (see in color on cover)

data that was mapped beneath the shot points actually came from other locations, up the dip of the formations.

Shooting 3-D is still expensive, prohibitively so for most reconnaissance surveys on land. But where it is used, it provides seismic data that fairly well depicts the three-dimensional nature of underground layers of rock. Interactive interpretation works with the 3-D data to give a picture of a three-dimensional subsurface.

FAULTS STILL AREN'T EASY

Picking faults has always been difficult. On paper it calls for folding sections, laying them side by side to compare them, etc.—all those things you do in trying to make your picking of faults consistent from line to line. Now there is interactive fault interpretation, and it helps a lot. But picking faults and relating them from one line to another still isn't easy. It is still often the hardest part of an interpretation.

There are some faults that are clear-cut on vertical sections, with noticeable throw, sharp breaks in reflections, and bands of reflections that can be correlated across the break. And there are faults, not necessarily the same ones, that are similarly clear on time slices. But there are also faults that are subtle, with little throw, unclear breaks, and with the reflections not correlatable across them. And there are faults with large throw that is difficult to determine by correlation.

Even after a fault is picked, it can be hard to visualize. Interpreters are seen explaining faults with their hands, like the "hangar flying" of pilots using their hands to describe their adventures.

A great deal of interpretation goes into drawing a fault on a section. Interpreters tend to make faults smooth, drawn as straight lines or long gentle curves. But looking at a section closely, you may get the impression that the fault has more of a zigzag shape. There is some support for the zigzag in our ordinary knowledge of how things break. Breaking a number of rock layers of different degrees of hardness, plasticity, brittleness, and thickness does not sound like a formula for a smooth cut. So probably the faults are more irregular than the interpretations show them to be. This tendency to smooth faults in seismic interpretation may derive in part from the way a geologist is forced to draw faults, with a fault seen as only one point on each well it cuts.

FAULTS IN 2-D

Connecting faults from line to line may be the worst problem there is in 2-D interpretation. Consider four lines that intersect to form a loop. Say that two of the lines, more or less parallel to each other, are each crossed by several faults. The faults are clearly visible on the sections, so there is no question of their existence or of where they cut the sections.

But then how do you map the faults? A fault on one line is likely to be the same fault as one of the faults on the other line, but which one? There are several on each line. There is often no way to tell which connects to which. But you have to draw the map some way, so you do all there is left to do—guess.

Interpretations are regularly based on these guesses. The maps show closures that depend on the faults. People know the faults aren't certain as drawn, but have no choice, other than making a different guess. Wells are drilled on the closures. A well may be cut by a fault, showing that the fault on the seismic map was in the wrong place. So the fault is redrawn to agree with the fault found in the well. This expensive trial-and-error process can go on for years.

Suppose a 2-D seismic line that runs north-south crosses a fault. The fault is clearly seen on the section as a break in a reflection. The fact that a reflection is broken means that the line intersects a fault at that point. If the reflection is deeper immediately north of the fault than it is immediately south, then the throw of the fault is down to the north. But wait, the fact that the line intersects the fault doesn't mean that it intersects it at a right angle. There is a range of angles that the fault could take. It can't be parallel to the line or nearly so, or it would appear nearly flat, and would look like a reflection from a formation. So the fault may be at an angle of, perhaps, anything from 45 degrees off the line's north-south alignment in one direction to 45 degrees in the other. That's about a 90-degree difference! The only clue to the direction is found in faults observed on other lines. If another section from a nearby more-or-less-parallel line has a fault that is also down to the northward, then the two indications may be of the same fault. That gives a direction for the fault trace. But suppose the second line has two faults on it. Which is the same as the one on the first line?

You can best handle the problem of matching a fault on one line with a fault on another by correlation of the faults themselves and nearby reflections. Systems that allow a concertina display can have several more or less parallel lines on the screen one above the other. Then you can correlate the fault breaks by eye. When you recognize two faults as being the same one, there is usually a way provided to change the color of one to the same color as the other, and for the computer to store them as one fault.

If your identification that two fault cuts are from the same fault is incorrect, then any determination of true dip and throw from several lines is meaningless—no, it's worse than meaningless, it's wrong.

. . . AND IN 3-D

In 3-D data, if a fault can be recognized on one line, it can be followed to another. And a section that crosses a fault at a right angle can be generated from 3-D data. So 3-D has solved some of the fault problems. But the others remain. They boil down to one question: How do you pick a fault?

RECOGNIZING FAULTS

With paper sections, you can interpret a fault by coloring the line of breaks in a section. You can correlate a reflection across the fault by "eyeball", just looking across. Or you can fold the section so as to conceal the breaks and match reflections on the two sides. Or you can cut the section (or a copy from the office copier) and overlay one part on the other. Once you get faults picked on another nearby section, you can place one section over the other and try to correlate them.

What does interactive interpretation offer to take the place of these techniques, or to improve on them? There are some quite effective methods.

A big problem with many faults is just seeing them well enough to pick with confidence. To make it easier to recognize a fault break, you can vary the horizontal-to-vertical proportions of a section. A great deal of horizontal compression makes a fault appear nearly vertical, so the breaks are lined up one above another, making the fault easy to recognize. But you can't pick very precisely on such a squeezed-up section. That's O.K., pick anyway, then expand the horizontal scale. Your picks expand with the section and give overall guidance; then you can improve the detail. That is one great advantage of interactive interpretation; you can pick in one mode, change the mode, and automatically see the picks in their correct places on the new display.

Another way of making it easier to recognize the breaks that indicate a fault on a section, and to correlate across it, is to display the section in various different color patterns and with various attributes shown. Try several color patterns; the fault may look clearer to you on one than another. And any tentative picks you have already made will show on all the displays, so you don't lose track of where you picked the fault. A velocity display may indicate a fault, with a velocity change from one side to the other. A layer on one side may have the same velocity as that layer on the other, or the layer may have a higher velocity on the downthrown side because of its greater overburden. A phase, frequency, or amplitude display may also have some distinctive features that are broken by the fault.

Since breaks in reflections are the main evidence of faulting, the more breaks a fault has the easier it is to recognize. So a display that has more things to be broken can be better for picking faults. For instance, a wiggle-VA section with both peaks and troughs filled in has twice as many strong features to appear broken as a similar section with only the peaks filled in. Also, a phase display, by making all reflections equally strong, has many events to be broken.

CORRELATING ACROSS FAULTS

For correlating across the fault, systems have one way or another of placing part of the section over another part. In one technique, some commands allow you to draw a polygon on the screen (any shape you want that is composed of straight lines) and to move a copy of that polygon around, like moving a piece cut out of a paper copy of the section. In some systems, picks that had been made on that part of the section are carried with the polygon (Fig. 8–2). In others, you may have to indicate just where the horizon is by the shape of the polygon, for instance by shaping it so its outline has a notch at that pick. Usually the best way to use the polygon for fault correlation is to form it more or less parallel to the fault, but set back some distance so you don't include part of the section that is distorted by the fault. Then move the polygon so it overlaps good data on the other side of the fault. Shift it up and down until you get a correlation. If the fault is a growth fault (so the two sides are not the same thickness) make it match at one reflection, pick the reflection on the now-adjoining part of the section, then shift the polygon so another horizon fits, and pick that horizon. Then you can abandon the polygon and extend your picks on both sides toward each other and the fault. One system also allows you to keep the polygon and add another part to it, in order to correlate across another fault.

Instead of a polygon of any number of sides, the system may use only a rectangle that may be moved around and slid up and down. The sliding may be in individual steps of one sample each, usually 4 ms.

There is an interesting alternative to the polygon. It is a magnifying

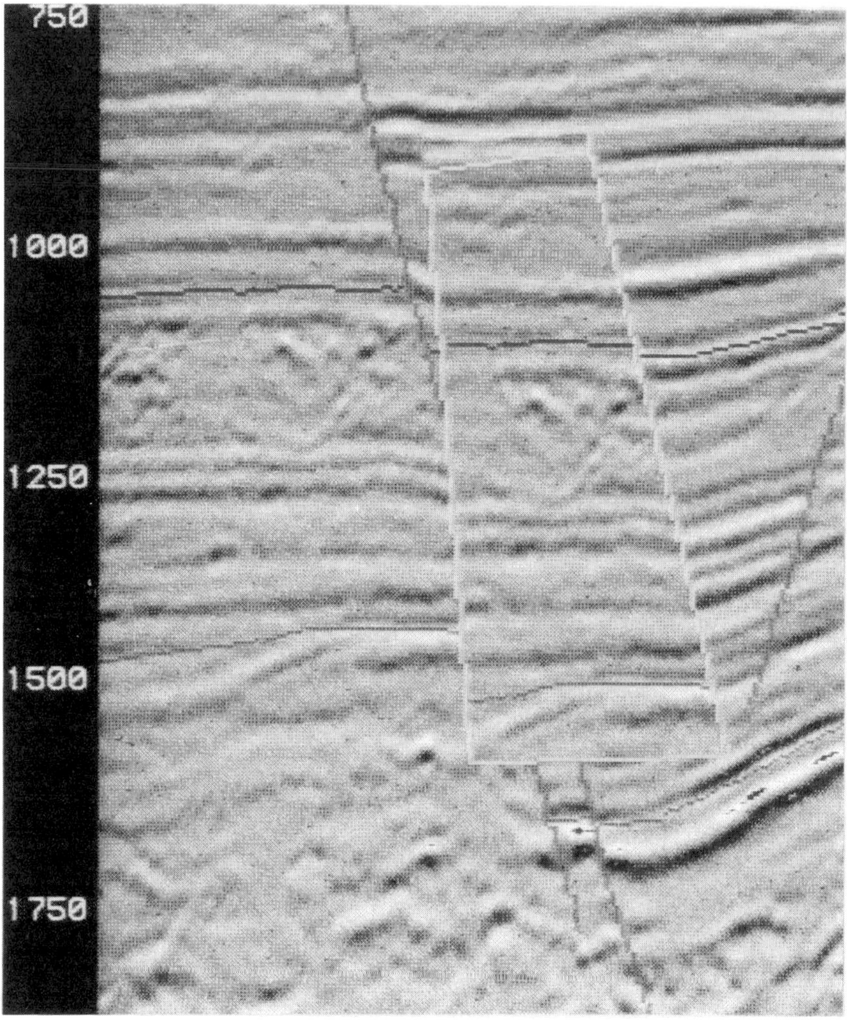

Geco Geophysical Company, Inc.

Fig. 8–2 Correlating polygon

function to display a portion of the section at a larger scale, in an immovable rectangle in the bottom part of the display. This enlarged part can be displayed in a different mode, for instance in wiggle-trace to see detail when the regular section is in variable-density mode. A second enlarged part can be shown in a rectangle adjacent to the first (Fig. 8–3). So two parts can be selected from opposite sides of a fault, and shown together at the bottom.

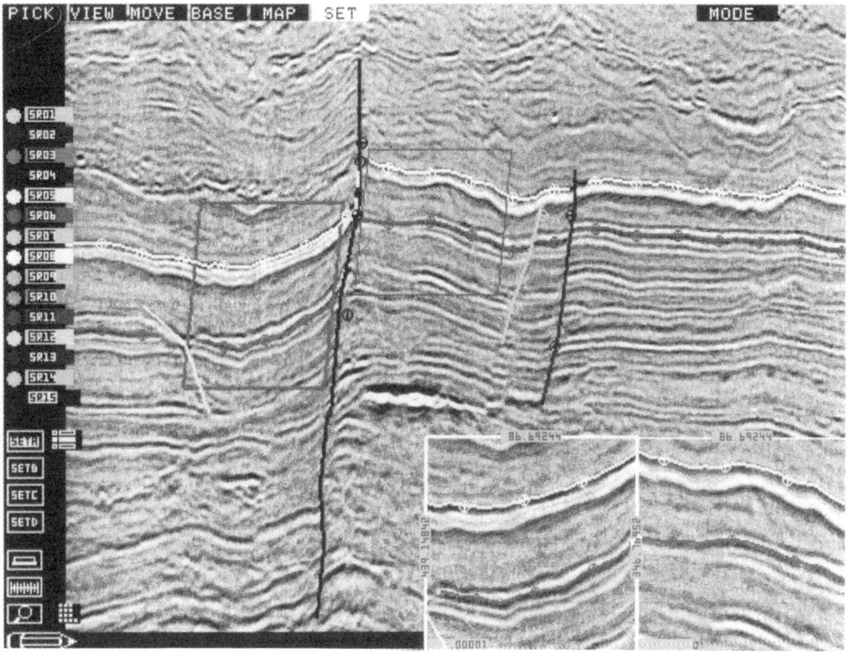

Graphic Science Corp.

Fig. 8–3 Square windows

Up where they came from are outlines of the areas that contributed them. These magnified boxes are sort of windows, and you can change the parts of the section displayed in them by changing the outlines of the areas they came from.

It is interesting to see the changes in the windows. Move a block on the section up and down, and the view in the window changes. Make one outline take in a taller area, and, to fit it in the window, the computer compresses it vertically. That enables you to stretch and squeeze one piece to allow for a growth fault that has made one side thicker than the other. If you want to parallel the fault, you can even push the top of the block to the side while leaving the bottom where it is, so its outline forms a rhombus. Then the magnified display distorts in the opposite direction to fit in its rectangle. Any picks you make on the magnified parts also appear on the regular section.

Or, instead of extracting and moving just part of the section, the display of the whole section is in two parts on the screen, treated like two sections joined at an intersection. If you move one part toward the other, the part of it containing the fault vanishes as though it had slid under the other section. Then you can shift one part up or down until a horizon matches, and pick across the break. If the fault is a growth fault, then you will need to shift for different horizons. In other systems, you can divide the section along a slanting line, for instance parallel to the fault, and then shift and pick (Fig. 8–4).

Some systems treat faults as horizons while others have special ways to handle them as faults. When they are considered horizons, they can be mapped like horizons (even though they are a lot steeper than most horizons). Some systems permit making more than one pick on a fault (or horizon) at one trace. That is useful for near-vertical faults that follow irregular paths on the sections, and for picking a number of faults before you have decided which ones they are.

A fault plane can be mapped like a horizon. From that map, true dip and the fault trace (strike) can be determined. In 3-D, a section can be displayed that crosses the fault trace perpendicularly, so it shows the maximum dip of the fault. If you wish, you can look at a whole series of these lines perpendicular to the fault.

With 3-D data, of course, you can make time slices. Such a display can be quite informative about faults. Where a section in the proper direction shows the dip angle and throw of a fault, a slice shows its horizontal direction. On a time slice, if you can see where a fault is on one line, there is usually no problem in continuing it to the next line. You can see it cutting across peaks and troughs. With some aspects of the fault clear on a vertical display and others on a horizontal one, a great deal of information can come from combining the two. So a combination of a slice and section joined at their intersection shows the fault from both viewpoints (Fig. 8–5). Then you can keep the same slice on the screen and display different sections with it, or hold a section and move down it with different slices. All this time you can be marking picks on both kinds of display, and changing them as your ideas change.

All this sounds like it makes faults easy to interpret. It helps a lot, but many of them still aren't easy. Reflection quality may be poor. Multiples may interfere. Sometimes the problem is in the nature of the fault. On a section a fault with only horizontal movement would not break horizons, and would be extremely hard to see. It would be between reflections, or would perhaps be a reflecting surface itself, indistinguishable from others. The idea of a horizontal fault sounds strange, but "sled-runner" thrust faults tend to flatten out at the bottom, with one rock layer sliding across another.

Similarly, faults that are almost parallel to the reflections on a slice are not noticeable because they don't break reflections. Faults that are at a large angle to the reflections, break those reflections, so they are visible as breaks. Even on a map, faults parallel to contours are hard to see. Because

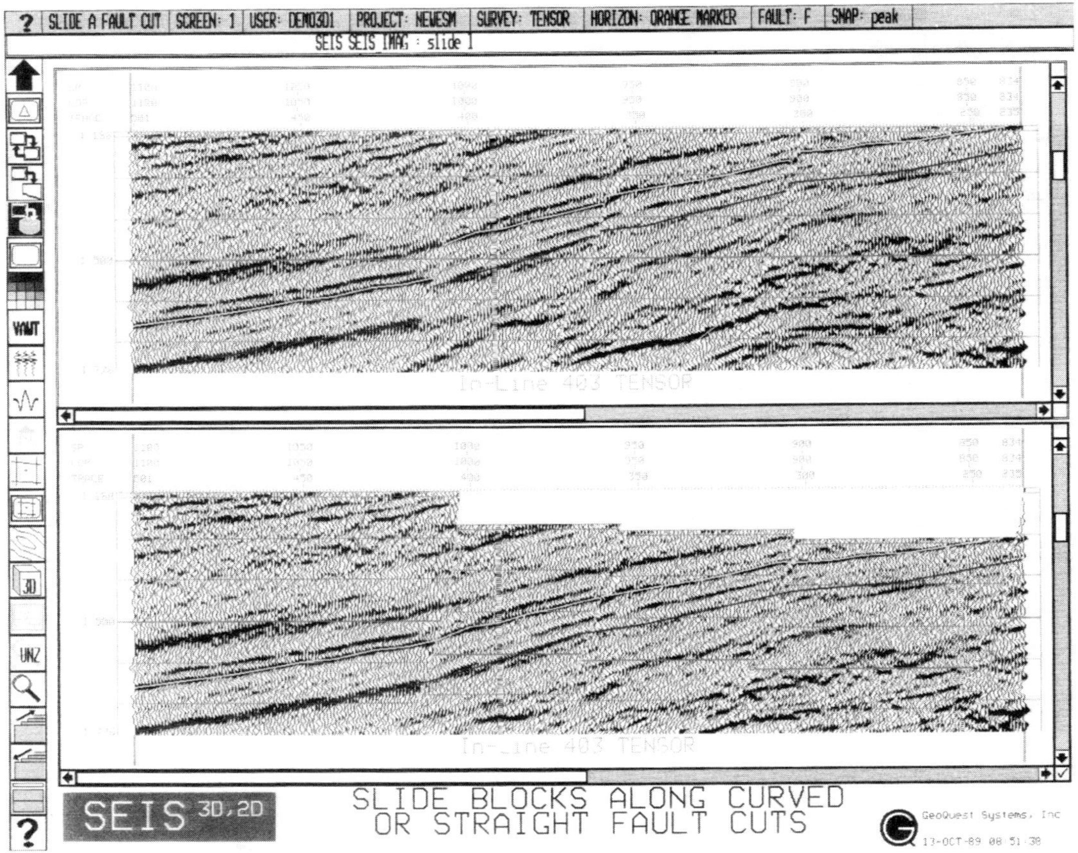

GeoQuest Systems, Inc.

Fig. 8–4 Section before and after fault shift

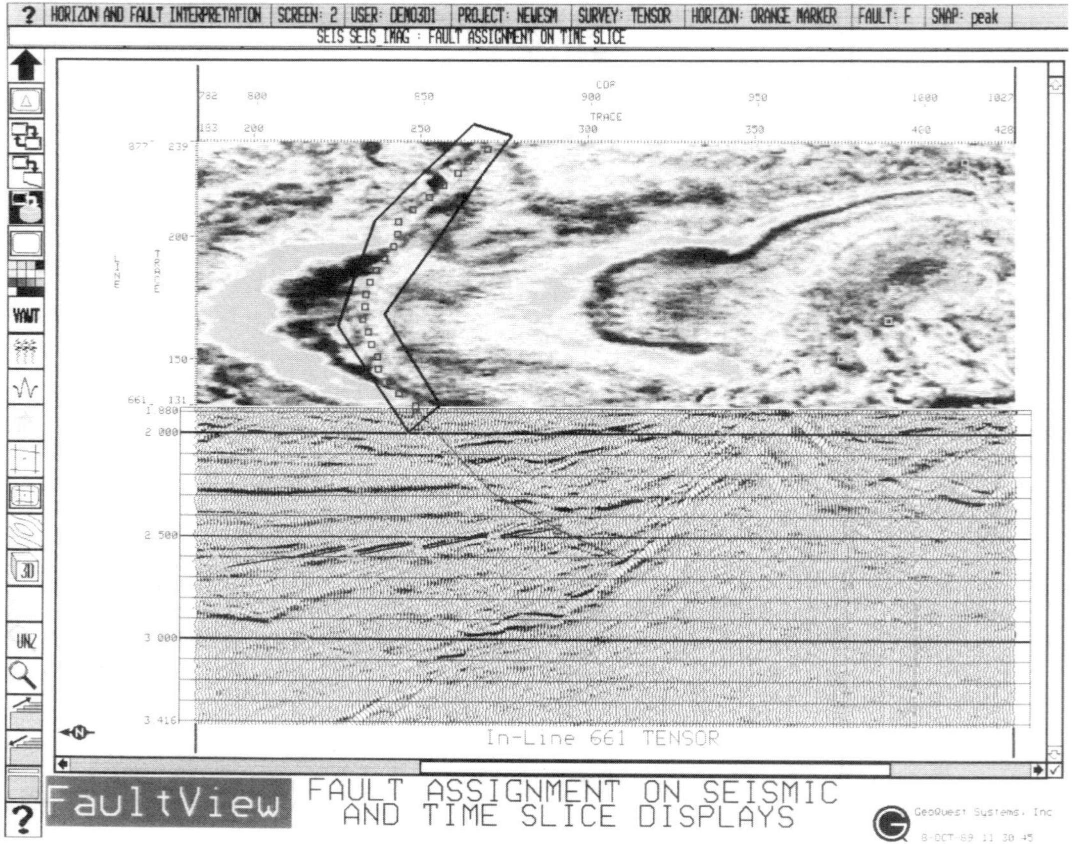

GeoQuest Systems, Inc.

Fig. 8–5 Fault on slice and section

of that problem, faults on maps are conventionally drawn in a different weight of line from contours.

If your system can pick automatically, not only along a line but from line to line, it may be able to work around the ends of faults.

FAULT DISPLAYS

Not only are faults difficult to pick, they are also difficult to visualize after they are picked. To help with the visualization, a system may display a three-dimensional pattern of faults. To make it, at first a fault is shown as a single line, the interpreted line of breaks in the reflections on a section. Then when you decide that it is the same as one on another section, you can connect them, and the fault plane will be shown. The representation of the plane is made up of parallel lines with gaps between them, so another fault plane behind that one can be seen through the gaps (Fig. 8–6). A separate color for each fault helps make the display clear. But if there are several faults in the area, they can overlap so much that it is still hard to distinguish the individual ones. To make it easier to see them separately, the display can be rotated. When you get the display turned so you see some faults end-on, you can separate them in your mind. The visualizing may cause you to change your mind about which fault connects with which. A system may have the ability to include mesh displays of horizons in with the faults.

The physical handling of faults involves several features in a program. When you are picking, there may be a break in the horizon either because there is a fault or for some other reason. The system can't tell the difference, so you have to supply that information. When you end a segment of picking, you can use a button to indicate that the break in the horizon is caused by a fault. Then the system holds that information in memory. If you have picked a fault, or pick one later, within a few traces of the break, the system can extend the horizon to the fault. You can make a similar indication when you start again on the other side of the fault, and the system can extend that side as well to the fault. Then it can determine the throw of the fault and the width of the gap it makes in the horizon. Later, with a number of fault breaks picked and indicated as such, you will be able to link some of them into a single fault. Having that, you or the system can draw an outline of the fault on the horizon. That outline is included in any contoured map of the horizon that the system produces.

FLATTENING

In interactive interpretation, an already-picked horizon can be flattened by giving a simple command. The line marking the pick becomes exactly flat, and the traces are moved up and down with it. If you aren't using an

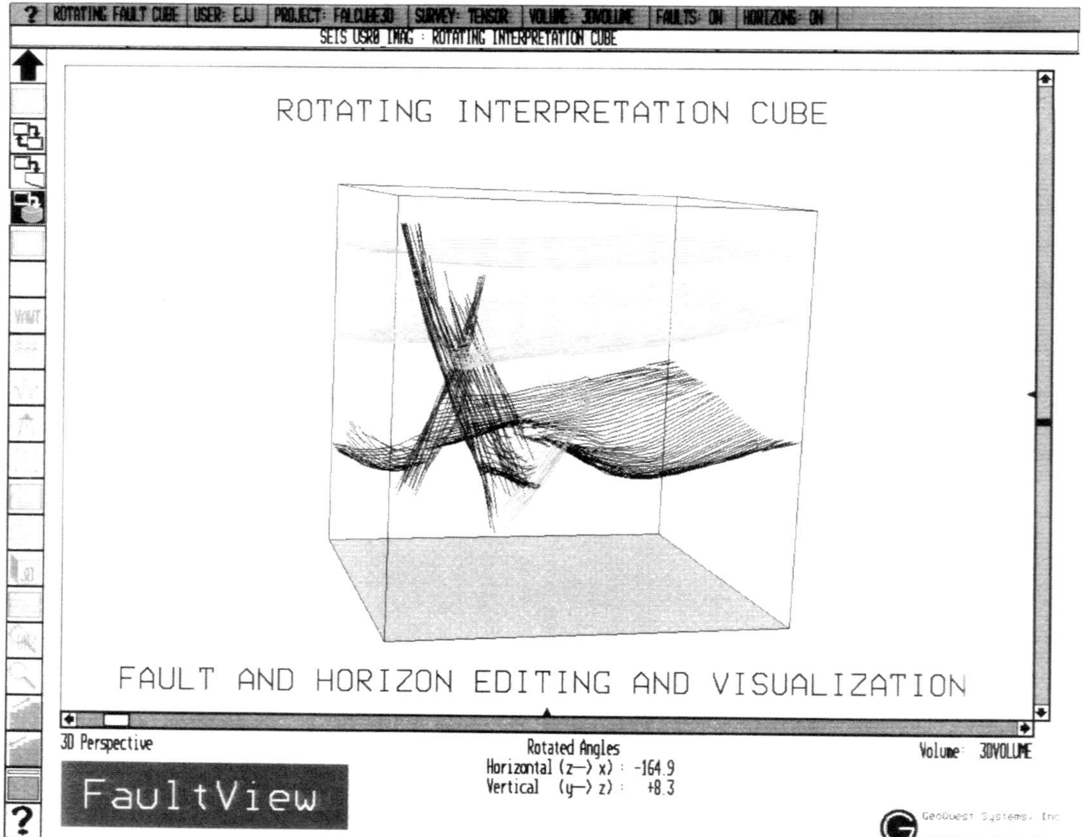

GeoQuest Systems, Inc.

Fig. 8–6 Fault display

interactive system, you can have flattening done by data processors. But like other functions, the ability to try it when you think of it, and see the results immediately, makes interactive flattening many times as useful.

When you give a command to flatten, you indicate the picked horizon to be made flat, maybe by pointing to it, then move the cursor to the time where you want it to be displayed. Traces move up or down the amount required to put all the points on that horizon at the specified time. That adjusts all the other reflections, moving them up or down and changing their dips by the same amount (Fig. 8–7).

The reason usually given for flattening a horizon is geological, to show paleostructure. Except in the shallow water near coasts, sediments are deposited approximately flat. Finely divided particles drift slowly down through the sea, and settle in a smooth layer, like a gentle snowfall. Irregularities are covered so the new top surface is smoother the way snow covers rocks and tin cans, first rounding the irregularities, then, as more falls, completely smoothing them out to form a flat surface. Later, the sediment becomes rock, and still later that rock layer is deformed and broken, so it is no longer flat. Flattening a horizon on a seismic section restores it to the attitude it had when deposited. When it has been flattened, all those layers that were laid down before it are in the positions they had when the horizon was deposited—flat or distorted by movement that occurred before the flattened horizon was laid down.

If you have picked a number of horizons, then you can flatten each in turn, preserve them on disk as still pictures, or make hardcopy of them. That gives you a geological history of deposition and tectonic distortion along that line. That is the geological reason for flattening, but there are also seismic reasons.

There can be near-surface problems that are not fully resolved in data processing. These include glacial drift, shallow coal or lignite formations, and permafrost. The problem is especially damaging if the prospects sought are very small or subtle. If there is a shallow strong reflection that is known to not have much dip, then flattening it is a reliable way around the near-surface problems. After flattening the shallow horizon, pick the deeper, more geologically significant horizons. The apparent structure on them represents a time interval between them and the shallow horizon. Or it can be considered as a sort of structure, with a known error, that of assuming that the shallow horizon is flat. That known error is preferable to the unknown error that remains after attempting to correct for erratic changes in thickness and velocity. An alternate way of using a shallow horizon for this purpose, but without flattening, is to use the system's calculating function to subtract the times of the shallow horizon from those of a deeper one, and display that as a time interval map.

Flattening can also improve your picking of the flattened horizon

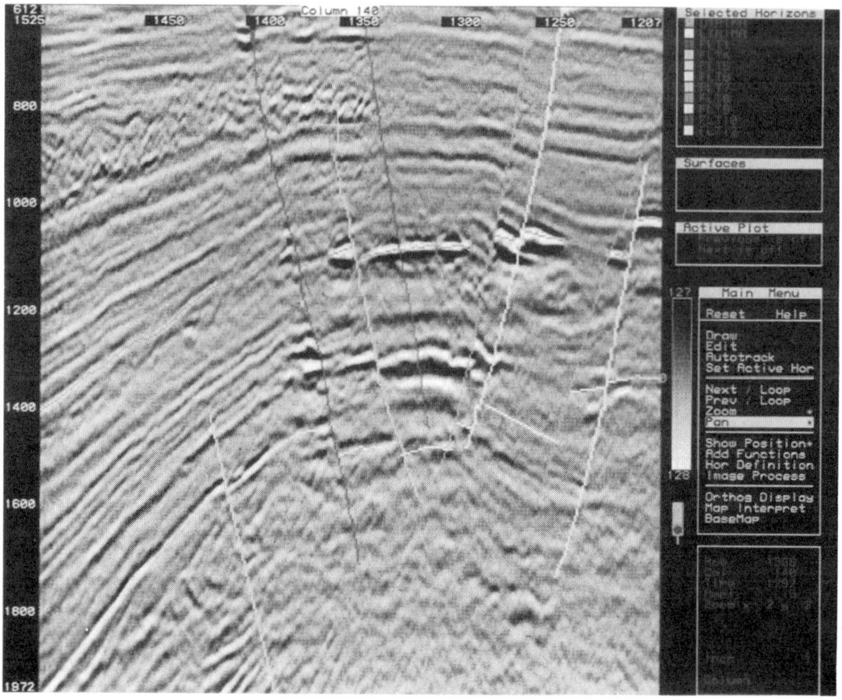

a Unflattened

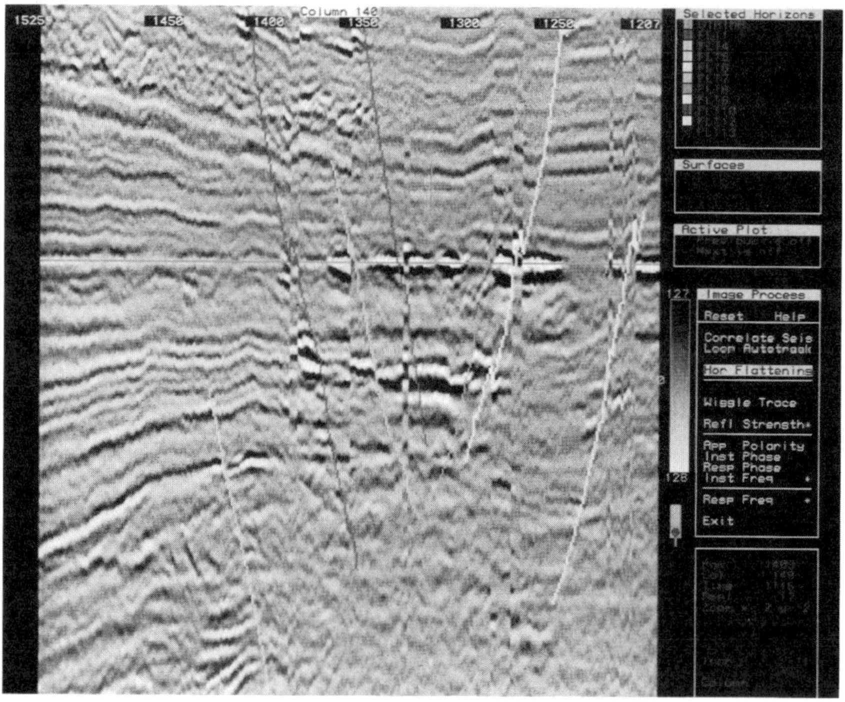

b Flattened

Geco Geophysical Company, Inc.

Fig. 8–7 Flattening

itself. On other wiggles near the pick, the section may not appear as flat as you expected, since slight inconsistencies between your pick and nearby events show up prominently. If other events, both above and below, form upward-pointing cusps, one above the other, then you must have picked too low at that point, and down-pointing cusps indicate that your pick was too high. Your system may allow you to modify your pick on the flattened section. If so, you can pick it parallel to the cusps and then tell the system to flatten again. Or you can pick one of the other horizons, flatten it, then re-pick and flatten the first one. If your system does not allow picking on the flattened section, you can call up the unflattened section, locate where your pick was too low or high, change the pick at that place, and then flatten the horizon again.

A variant on flattening the shallow horizon is to flatten any horizon that looks fairly good, and use the differences between it and any others to indicate the difficult-to-recognize structure.

Another seismic reason for flattening is to assist picking in areas of strong tectonics. Where there is steep dip or considerable faulting, it may be difficult to pick some horizons because of the distortion. You can measure a horizon's time below a strong reflection, but that is laborious, and the horizon you are picking is still distorted. So if you first pick a good, strong reflection (not necessarily a shallow one this time but one as close to the one you are interested in as possible), and flatten that strong reflection, you have made your horizon smoother and therefore easier to pick. After picking it, you can then display the section not flattened, and see the structure of the horizon you are interested in.

The same technique applies to a section with poor data. With it, the best approach is to pick the best horizon on the section. That is probably, but not necessarily, a shallow one. Then flatten that horizon, and pick the second best. After that, depending on which turns out to work better, you may want to flatten the horizon you picked most recently, or just leave the best one flat, as you pick more horizons.

Some systems flatten only the traces on which the horizon to flatten is picked, and don't do anything to the section in the gaps where the horizon was not picked. Others automatically interpolate across the gaps in order to flatten the section without leaving gaps. If yours is the kind that leaves gaps, you may on some occasions wish it could do the interpolating.

That isn't difficult. Just do your own interpolating across the gaps by using the rubber-band function to join picks on opposite sides of the gap, and then flatten. The computer sees the interpolations as picks, and produces a flattened section without gaps. Of course, that adds the interpolations to your picks, and after you have got the section flattened you may not want the interpolations any more. O.K., then don't bridge across the gaps in your main interpretation. Instead, start another horizon, with a different name, and pick it exactly where the first one was picked. Then interpolate on that horizon, and flatten it. If you don't want that interpolated horizon later, you can delete it.

The system may allow you to flatten a horizon over an entire 3-D area. That saves you the trouble of flattening each line individually, and gives you the benefits of flattening for the area instead of just on one line. Then you can call up lines one after another and see other horizons as they were when the flattened one was deposited, and pick them with tectonics removed. Making a time slice at the time of the flattened horizon gives you a horizon slice, with its advantages in showing you differences in amplitude and other characteristics of that horizon. The system may also let you make time slices at other times, so they are parallel to a flattened surface. You can use them to help you decide whether to pick those reflections. You may also be able to show a cube or other exotic display with a horizon flattened.

STRUCTURE INTERPRETATION

Structural interpretation is a matter of determining the shapes of rock layers. It is the older way of interpreting seismic data, and is now of special importance because some of the newer kinds of interpretation depend on correct interpretation of structure.

Structural interpretation is concerned primarily with the times of reflections. Usually the continuity of the reflections permits picking along a reflection. But where formations are broken by faults, or reflections are interrupted by poor data, or part of the line was not shot because of terrain or permit problems, parts of the reflection must be correlated. Besides the general convenience in handling data, what does interactive interpretation do for these processes?

Structure has conventionally been interpreted on amplitude sections. But a display of instantaneous phase can make it easier to see continuity. A phase display does not show amplitude, so your eye is not confused by changes in reflection strength. As long as the reflection isn't interrupted, it can be followed more easily on a phase display than on an amplitude display. The absence of amplitude information keeps you from seeing changes in reflection character, but when you come to a break in the reflection, you can switch to an amplitude display to correlate across the gap by eye. A wiggle-VA display may be good for correlation, since you can see the shapes of the wiggles, and see how much alike they are across the break. The picks you made on the phase display are reproduced on this display, so you can see exactly where you had picked. Other types of display may help the correlation. Velocities will help identify a reflection on the other side of a break. Similarly, frequency patterns may contribute to recognizing the reflection. After crossing the gap, you might go back to a phase display. Again the picks you have made on other displays are shown on this new display.

With any of the types of display, you can simplify getting across a break by omitting the bad part of the section. You can join good-reflection parts from the two sides of the break, and then slide one part up and down until similar-appearing reflection characters meet.

Faults can be made easier to detect by compressing a section horizontally, so they look more definite and more vertical. The fault-connecting functions make it easier to connect fault indications from line to line of 2-D shooting. Perspective displays of faults let you see how reasonable they look.

And, of course, the automatic timing and plotting of picks give you quick access to a structure map.

IN THIS POST-STRUCTURAL AGE . . .

We hear about the coming (or already here, depending on who you listen to) Post-Industrial Age, a time when services are more important than manufacturing. Similarly, and maybe easier to get a feel for since it's in the past, the Industrial Age followed the Age of Agriculture, and so was a Post-Agricultural Age. But does that mean agriculture went into decline in the Industrial Age? Gosh no, it became more efficient than ever. And we depend on it for all our food like we always did. This Post-Something just means that the something isn't any longer the highest technology of the time. It may still be necessary, but it can to some extent be taken for granted, while the new effort is on something else.

In that sense seismic interpretation is in a Post-Structural Age. One of the new ways to find oil is by seismic stratigraphy. There are still structures to be found, but most of the giant fields of the future will probably be stratigraphic. Fortunately, seismic data has become so good that it can now be used to find stratigraphic traps. Does that mean interpreters can forget about structure, that if they get the stratigraphy right the structure doesn't matter?

Well, no. It would be like saying that in this industrial age, farming no longer matters. Industry is possible partly BECAUSE farming is efficient. And in seismic work, most stratigraphic traps depend on structure. In most cases the structure must be interpreted reasonably correctly for the stratigraphic mapping to be useful.

A pinchout traps oil only if the formation thins in the updip direction. How much oil would it trap if tectonics had tilted it downward?

A reef holds oil by being a porous blob that sticks up into impermeable rock. If the reef is tilted so the oil spills out, it is not a trap.

For a horizon slice to have meaning, the structure must be interpreted correctly first. A plain time slice doesn't depend on interpretation, it's just a sampling at a certain time. But a horizon slice is made by first picking a reflection, then flattening that pick, and then displaying a slice at the time on which it is flattened. The main usefulness of a horizon slice is to show how some attribute of that horizon varies from place to place. Suppose you had picked a horizon, but in a poor-data part of the area you had mispicked, jumping a leg to the next event. Maybe there was a fault, and you mapped it but correlated across it wrong. Then you made a horizon slice, displayed in colors to show changes of amplitude. How interesting, there is a change in amplitude in crossing the fault! You start thinking of geological reasons for the change to occur at the fault. Could it be that the fault existed as a low scarp when the formation was being deposited, and affected the sorting of grains? Better stop right there and see if there is a simpler solution, like incorrect structure mapping.

3-D AND THE RESERVOIR

Exploration has been the main use for seismic investigations since the first seismic crews. For most of this time, there was not much done seismically to aid development of already-discovered oil and gas fields. There are several reasons for that: seismic data wasn't very precise, the data wasn't obtained in much detail, and people had the feeling that once a field was discovered it could be produced without additional seismic input.

Seismic work is now increasingly being done to assist field development. Seismic data has become more precise and reliable, and 3-D shooting can provide detailed coverage of a field. A decline in the price of oil has led companies to concentrate on the less-risky, shorter-term ways of obtaining oil—developing fields that have already been found rather than looking for new ones. In the U.S. the amount of seismic work in development has become about equal to that done in exploration. This has happened partly because of an increase in the development use, but also because of a decline in exploration activity.

Although in exploration you usually work one area and then another, in development you may be assigned to a single field for some time, so you become an expert in it. Also, in exploration you tend to work with landmen in addition to geophysicists and geologists. In development, engineers take the place of landmen. Exploration is free-wheeling, calling for imagination and optimism. Development is so constrained by data that it must be done more meticulously, but it too calls for imagination.

DEVELOPMENT—NEED AND OPPORTUNITY

Two factors combine to bring about this increased use of seismic reservoir work—a need for it and the opportunity to obtain it.

THE NEED

Development requires detailed information, much more than exploration does. Just finding something doesn't call for the precision that is necessary to describe it. In exploration, lines are shot to discover a possibly productive feature. If one is found, then more lines are shot, maybe in several stages, perhaps including 3-D, to determine the general shape and size of the feature.

Then instead of doing more seismic exploration, the company can drill one or more wells to find out if there is oil in the prospect. If the feature doesn't produce, the company probably won't have any reason to spend more money on it. But if it does produce, a great deal of money will be used to develop it. Locations for wells and platforms determine how completely the field can be drained, and how many wells must be drilled to drain it. Later, injection wells must be positioned so they will be most effective at driving the oil and gas to the producing wells. Drainage and the various kinds of flooding can be monitored to see the effects they have had on the reservoir. There is a great need for good seismic data to help guide this development.

THE OPPORTUNITY

Along with the extra need for detail is additional opportunity to obtain it. A 3-D survey can be shot before a lot of equipment is installed in the field; wells that will be drilled in the field will provide additional data; there will be time to make a thorough study of the field.

As a field is developed, drilling rigs on land or platforms at sea will be set up; pipes will carry oil, water, steam, around the area; trucks or boats will come and go; scrap pipe or buoys and anchor chains will abound. Seismic work will become difficult, from the viewpoints of both interference from the noise of machines and the possibility of damaging production equipment. The best time to get a good thorough seismic coverage of the area is after it is discovered but before the installations are set up. This is the one chance to get all the coverage you think you might need, and to get it with a close enough trace spacing and line spacing for possible future needs. It is not a time for cutting down the program to save money. The real saving will come from improved development of the field.

If the exploration stage included a 3-D survey, the survey may need to be extended to include what is thought to be the entire field and enough peripheral area to allow the field to be related to normal geology in the vicinity. The extension should include places where injection wells might later be drilled, or where the field might be extended. If there is no existing 3-D survey over the field, then one can be shot to cover all the appropriate area.

Although exploration must usually be done without much well data to help with either identifying horizons or determining meanings of seismic features, development seismic data just about always can be calibrated with wells. For reservoir studies, you will have more time, more money, and more wells to use for calibration. Those advantages let you make the really careful investigations that you often think you'd like to make in exploration. One of the keys to such a long-duration study is the opportunity to keep trying to get closer to the correct answer. You interpret seismic data, incorporating well data (Fig. 8–8). Production history adds new information, so you re-interpret to be consistent with it. More wells are drilled, and you again adjust the interpretation. The repeated adjustment keeps making your work better and better. You will probably never cease to be surprised by new data, but the surprises will be smaller ones as time passes. Each oil field has its own set of characteristics. As you work on a field, your

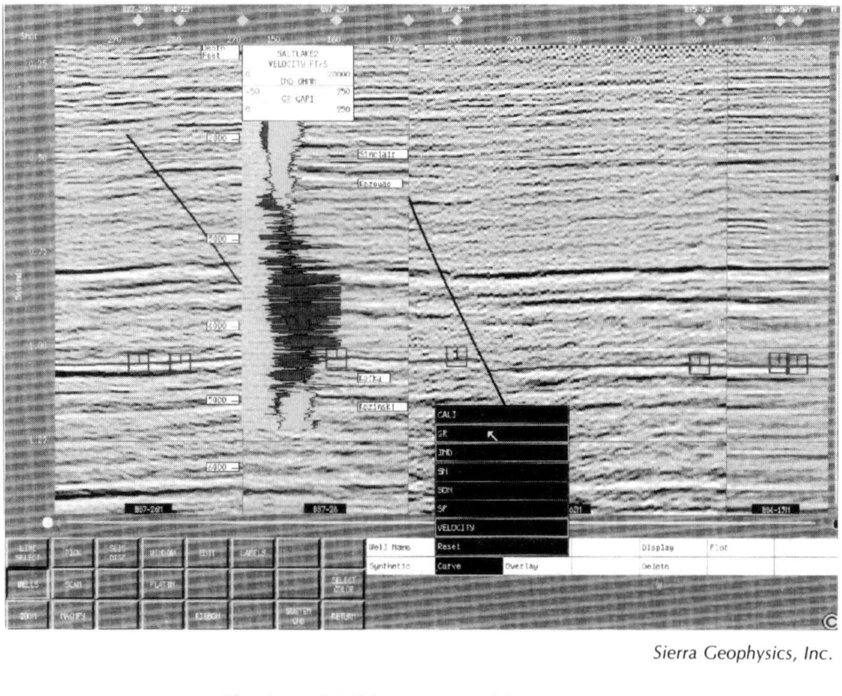

Sierra Geophysics, Inc.

Fig. 8–8 Well logs inserted in section

experience makes you better at interpreting that particular field. And that in turn helps in making decisions.

Both 3-D data and interactive workstations are vital to the combination of seismic data with well data. Shooting 3-D gives you seismic data at all the wells. It allows you to make lines through collections of wells. It permits you to have horizon slices to study formations in detail. Interactive systems enable you to call up the different kinds of information quickly and to make comparisons easily.

A different use of seismic workstations is as extra-powerful geological workstations. A seismic system, devised to handle the mass of data in a 3-D survey, is ideal for a geologist to use on a large oil field with a great many wells. The system can be used without modification, just treating individual wells and groups of wells as though they were shot points and seismic lines. The techniques for using such functions as picking horizons and faults, flattening, and subtracting one horizon from another, are as easy and effective for wells as for seismic data.

You can display on a section the wells that are on or near the line. Or, better, you can form an irregular line that goes through the wells. Well tops and production data from the wells can allow you to determine the meanings of features of the seismic data. Some amplitude anomalies may occur where there is gas in the wells, and others may not correlate at all with production. Characteristics like that usually do not apply the same way in other areas.

You can display well logs on a section, with producing zones marked on the logs. Then you can vary the display of the section, and look for similarities between section and logs or between section and producing zones. The variations of the section can be simple, like changing the color arrangement to enhance different amplitudes on a conventional section. Or you can try various attribute displays. You may find useful information in the length of some specific time interval, or a difference in amplitudes of two horizons, or an average amplitude between two horizons. Any correlation you obtain can be checked with other lines and other wells, and may help guide drilling and production.

The general structural shape of the field is probably known from the exploration shooting. As development wells are drilled, they give more precise data at their locations. You can fit seismic data to the wells, and get information on the structure between wells. The availability of plentiful well data makes it possible to obtain really good velocity information, so good that mapping in depth can be reliable. That is generally not possible in exploration work. At each well, you have a check on the velocity and on the identification of horizons. By adjusting at those points, you can map structure much better than you could before the wells were drilled.

Velocity information from inversion of traces into synthetic sonic logs gives a good idea of interval velocities. Velocity in turn can furnish information on porosity. Like structure, velocity can be adjusted to fit at each well location, or at each well in which a sonic log was run covering the levels of interest. With a number of wells, the porosity information can be fairly reliable. Combined with information about the field, porosity may translate into volume of oil and its distribution in the field. It may make it possible to drill in the best places to get the oil from the field.

Shear-wave investigations can be made with three-component geophones, that receive P, SV, and SH waves. A system may store both primary- and shear-wave data so they can be worked together (Fig. 8–9). You can expect to one day be routinely calling up and using those different kinds of data on your system. They are useful for investigating lithology, including yielding information on bright spots and fracturing.

Bright spots in ordinary P-wave sections may be caused by gas or by slower-velocity rock, but the ones caused by gas do not appear on S-wave sections. A comparison of the two enables you to identify gas in formations.

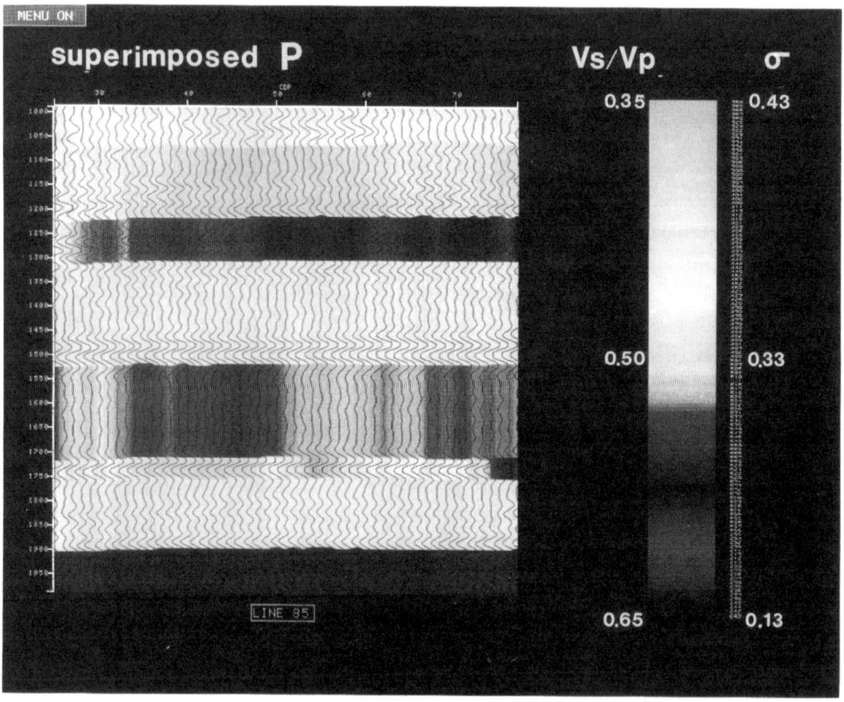

Petrosystems

Fig. 8–9 Ratio of shear to primary velocity overlaid on conventional section

Shear waves can be used to determine the direction of fracturing in formations. Faulting and the lesser faults that are called fracturing provide channels for oil to travel to wells. The pattern of broken rock can be determined with increasing accuracy as the field is developed, by seismic techniques with repeated checks from other types of information, particularly engineering data.

Time of shooting can be another variable. Development seismic work can be done repeatedly in the same area, to monitor changes caused by production and by secondary-recovery techniques like steam flood, water flood, fire flood, and gas injection. Steam and fire raise the temperature of the formation, and sound goes more slowly through hotter rocks. The advancing front of such a flood can be monitored by determining interval velocities in a formation from shooting at different times. Ray tracing can be used to analyze the interval velocities between horizons above and below the heated zone. The gases may also form bright spots that grow larger and maybe brighter as the flooding progresses. This later shooting may be done with the geophones buried permanently, so re-shooting can come close to duplicating earlier source-receiver patterns. It has the disadvantages of having to be done where there is much equipment to be worked around, and where machines are creating noise.

With wells available, VSPs, having the geophones down a well and the sources on the surface, can be shot. An inverted VSP has the source in a well and geophones on the ground. Crosshole shooting, with sources in one well and geophones in another, uses largely horizontal travel paths. Shooting in some of these different directions can be combined in seismic tomography. Rock type can be determined by combining VSP, shear-wave shooting, or AVO (amplitude vs. offset) with data from well logs, cuttings, and production history.

The two factors in development seismic investigations fit together well—**discovered fields need especially good seismic interpretation; the information generated in the fields makes it possible.**

GETTING THE PEOPLE TOGETHER

A workstation needs to have access to a lot of data, like geographical information, well logs, production histories. Such an assortment of information is handled best in a database, so any part of it can be called up quickly.

When it is thus available, it would be a waste of opportunity to not use it fully. But some of the information is input from other specialists—surveyors, landmen, geologists, engineers—and is constantly changing. Companies acquire and surrender leases, geologists re-correlate logs, production histories produce new information. A seismic interpreter needs to have a database that is up to date. The other specialists need input from changing seismic interpretations. The ideal situation is for all to work together, drawing on one another's results, prompting one another to reconsider conclusions, in general allowing all to do better work.

Even without databases and workstations, companies have for years been trying to get geologists and geophysicists to work more closely together, even though the ways the companies are organized may contribute to the separation. Now, with this wonderful tool available, the need is for not just those two specialists, but all affected ones, to use the tool to gain better knowledge of regions, prospects, and producing fields. If all the specialties can put data into the system, take results from it (Fig. 8–10), and work together on problems, the company will find and produce more oil.

But that goal of thorough cooperation is in general still some way from being achieved. The specialists do usually work separately. One person may often wonder why another specialist got some strange result, but rather than comparing notes, may shrug and go on without the better information that a discussion would have led to. Even seismic work is separated into the distinct specialties of acquisition, processing, and interpretation, and there often isn't enough communication among specialists in those fields.

A shared database makes all the kinds of information available to each specialist. If one person knew enough about all the specialties, that person could conceivably handle the data alone. But even a person who was skilled in all of them could not stay familiar with the current status of each. Cooperation among specialists is better—calling up one another's data, investigating, talking to each other, arguing, and coming to better conclusions (Fig. 8–11).

DON'T WAIT UNTIL EVERYTHING IS PERFECT

Suppose, when you are ready to start using the system, not everything is ready for you. Maybe you are going to combine several 2-D areas, and one of them isn't loaded yet. Or maybe the seismic data of a 3-D area is loaded, but the wells aren't. It would be easy to wait, to say "I'm waiting for the last set of lines" or "I can't get started; my wells aren't in the system". But, unless your project is one without any time pressure, that wouldn't be practical. Seismic interpretation is done to find oil and gas, or to efficiently develop oil and gas that has already been found. It is not an academic exercise. You can begin without all the information.

Working on paper sections in newly shot 2-D areas, people often start interpreting as soon as a few of the lines are processed. It would be more efficient in a way to wait until they were all done, and then sit down and interpret the area as a whole, but that takes too long. If you start as soon as possible, you get useful information sooner, and also finish

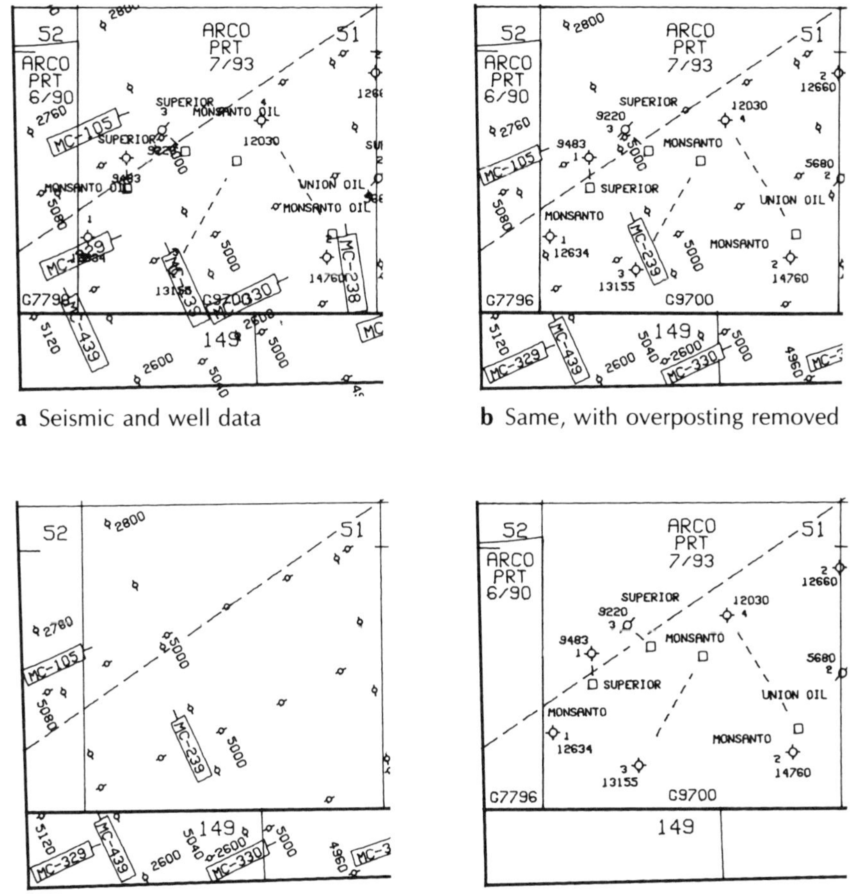

a Seismic and well data

b Same, with overposting removed

c Seismic data extracted from b

d Well data extracted from b

Energy Graphics and Jebco Seismic, Inc.

Fig. 8–10 Information from database

the project sooner. That matters in oil exploration and development. A well may be completed sooner than expected, so the rig that drilled it is suddenly ready to go to a new location. In that case, you may need all the early information you can get, to help select that location. It's not pleasant having expensive rig-standby fees piling up, waiting for you to complete an interpretation.

In the 2-D example, it makes sense to work on the lines you have loaded. You can see what the main features of the area are, pick horizons,

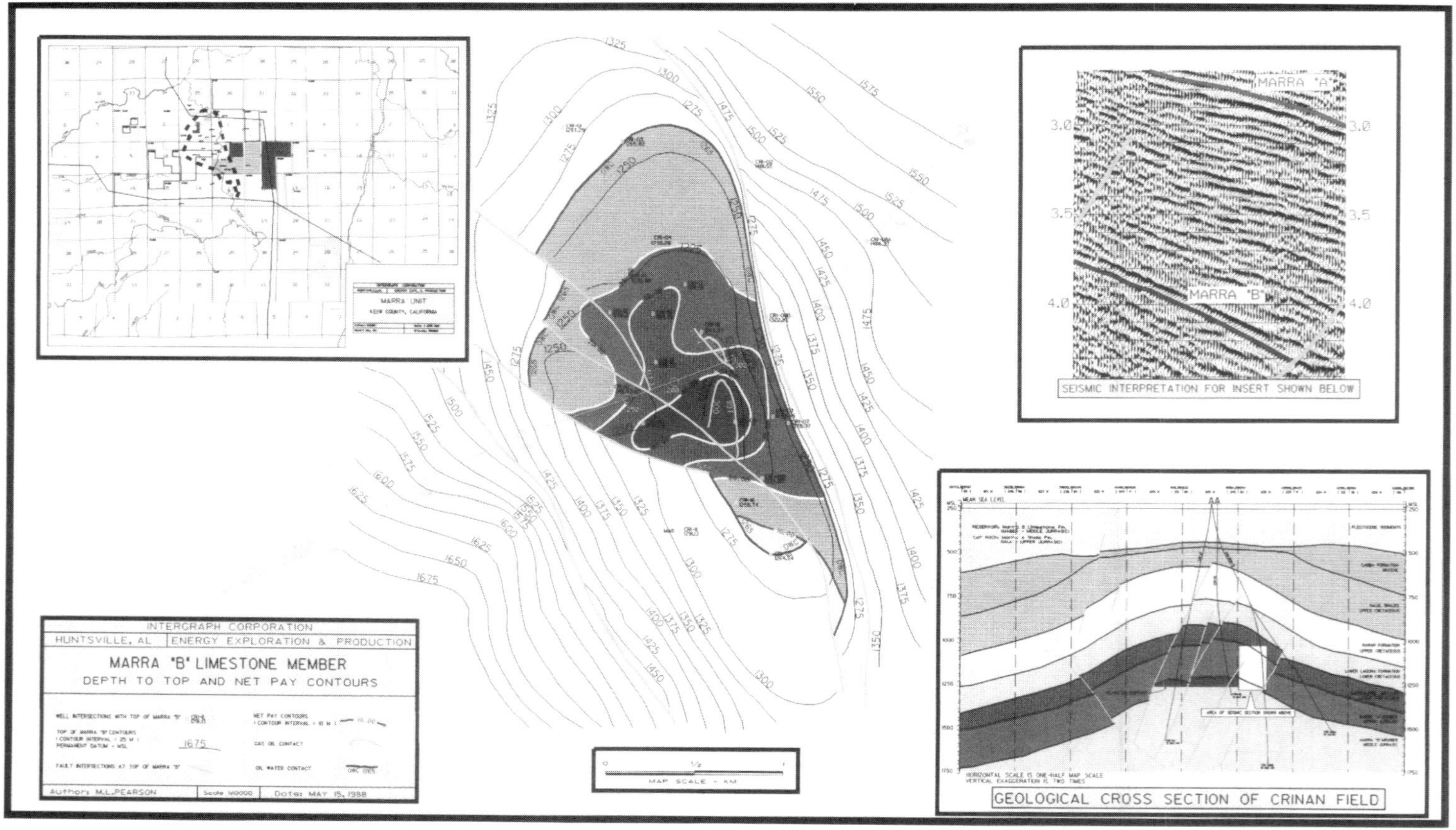

Intergraph Corporation

Fig. 8–11 Several kinds of data

experiment with different kinds of display. Later, when you get the rest of the area loaded, working it in doesn't take nearly as long as working the whole area. You may lose some ground, maybe having to change some picks you've made, but you are already familiar with the area and its problems. You can re-work part of it faster than you worked that part the first time.

In doing interactive interpretation on a 3-D area, you do have to wait until the seismic data is all put together and loaded in the system. If there are wells in the area, but you don't have the well data loaded, you don't have to wait for it. You can first check the area over, see what the significant reflections are, and how difficult they might be to pick. You can pick some horizons, even though they aren't yet tied to well data. You can work out the faults; that's a large enough task to keep you occupied for some time. To get a start at identifying horizons, you can plot formation tops from wells on a paper shot point map, and determine their locations in terms of line number and shot point number (for deviated wells, each horizon will be at a different location). Then you can call up those lines on the screen and mark the points on the sections. You can do that using an annotation function or just by initiating a new horizon and putting the marks in as though they were picks. That is good enough to identify a reflection with a formation top, and to see how the reflection ties from well to well. Later, when you get the well data loaded, you can check the steps you did manually, and can do some of the more sophisticated things, like overlaying synthetic seismograms on seismic sections.

You can get by, at least temporarily, without a number of kinds of information. There isn't any need to just wait.

9

Powering Down

An interactive system is a wonderful aid to interpretation. It has a powerful computer and good software. But people have been interpreting for years without all that. Is interpreting really any better on one of these fancy machines than with a skilled interpreter working carefully on paper sections?

Yes, it can be a lot better. The book is full of advantages of an interactive system. And to keep things in perspective, we might as well cover its disadvantages too. It does have disadvantages—everything does. Some have already been mentioned, particularly in the segment "Computer Problems".

NEGATIVES

Most of the disadvantages of an interactive system are caused by its use of electronic equipment.

First you have to have a system. A complete system may be the size of a desk. Or it can be so large as to include a computer that takes up one side of a room. A system can be a large investment. You have to foresee having enough interpretation to justify buying or leasing it. To get your company to acquire a system, you have to put together convincing reasons for getting it.

It has frailties that a hand method doesn't have. It depends on a constant power supply. If there is a power failure in the building, the system will quit working until power is back on. And depending on the safeguards that are built into it, the power loss may make it lose some of your most recent interpretation. A thunderstorm may also have a bad effect on data and the machine. A standby generator may be needed if power failures are frequent.

The computer is sensitive to temperature and humidity. It needs air conditioning to work properly. Either the building must have steady, reliable air conditioning, or your company must provide its own. And the air conditioner too needs that steady supply of electrical power, like the computer does.

Interactive interpretation systems aren't silent. You have to put up with the constant hum of their fans and disk drives. Some systems allow you to avoid the sound during your interpretation, by having the computer in one room and your terminal in another. If your system works off a mainframe, you are normally in another room.

Systems also have you looking at glowing screens. At one time monitors produced a flicker that tired the eyes after a few hours. But some or all present-day systems have monitors that do not flicker, so that problem doesn't seem to exist any more. The glare is still there, though. It doesn't feel bothersome, but one study has suggested that long-time viewing of monitors may harm the eye's ability to focus sharply. The study was probably conducted on people who were looking at text, which has greater contrast than seismic sections. But it might be a good idea to not have uninterrupted sessions of many hours at a time, or if you do interpret in long sessions, it might be sensible to turn down the brightness and contrast on the screens (with controls like those on a television set) to reduce the glare.

The screens are small. You can't view a long section or a large map on them easily. The limitation is in the number of pixels the system will handle, so getting a larger screen would just provide larger pixels, not more of them, and you wouldn't see any more detail.

The system must be in working order. If it quits, you probably can't fix it. An expert will have to come from somewhere and take things apart for what seems to an impatient interpreter to be a long time. Then the expert will have to test it, and maybe take it apart again. Even when there is no problem with the system, your interpretation will sometimes be interrupted while routine maintenance is performed. The maintenance is necessary to prevent breakdowns, but it's an interruption anyway.

Although you may already know how to interpret seismic data, the system has a whole new set of skills for you to learn. And if you learn the commands to do some particular thing with your data and then don't do that for some time, you may find yourself bungling the commands when you start using them again.

Although there are safeguards against it, it is possible for you to lose the interpretation you have done, especially if you haven't made backups.

A company won't necessarily have a lot of systems lying around, so you may be limited in the times you can use it. You and the other interpreters will have to take turns using the equipment.

If you solve some of the problems of high demand for a system by deciding to work at unusual times like evenings and weekends, then an arrangement may have to be made with the building management to leave the air conditioning on. That may require approval from someone in your company to pay for the extra service. And even after an arrangement has been made, the building people may forget and leave you without cooling. You may be uncomfortable, and the computer may start showing strange displays, and then quit working altogether.

Working odd hours also can lose you the opportunity to interact with the people in the office. Alone with the system at night, you can't get assistance if you need help with the computer, and you can't compare thoughts with geologists and geophysicists.

The system is installed in a fixed location. You can't take it into the conference room for a meeting. If you want to show something in the area to your boss, either the boss has to come in and look at the screen, or you have to make a hardcopy to take where the boss is. If you are away from the system and the boss asks a question you hadn't anticipated, then you may have to go back to the computer for another printout. You can't work on it at all when you go out of town to visit a field crew, or to attend a meeting.

You may encounter resistance on the part of other people to maps and other results of the interpretation that are made by a computer. Then you will have to have some of the work done over by drafting.

Interactive systems have huge storage capacities, but the amounts of data to be stored are growing rapidly. You may overload the system, and have to break the area down into segments and work them separately.

If your system uses the company's mainframe or minicomputer there probably isn't much chance of running out of memory, but then you are timesharing with all the other company users of the computer. Whenever there is an extra load on the computer, then for all the users its speed of operation slows down. This can happen when other geophysicists are interpreting at other terminals, or when the data processors are using a power-hungry program like migration before stack, or at the end of any month when the accountants have their heaviest use of the computer.

The program that runs the system can, like other programs, have "bugs", flaws in the way it is written. Several of these problems give you an excellent chance to use the ever-popular expression, "My computer is down".

POSITIVES

Like the disadvantages, the advantages of systems result from their use of electronic equipment, particularly in the power and convenience of computers (Fig. 9–1).

The fact that the systems provide a way to handle the vast quantities of data in a 3-D survey is one of their big advantages. The data would totally bog down a person trying to do it on paper. You would have to

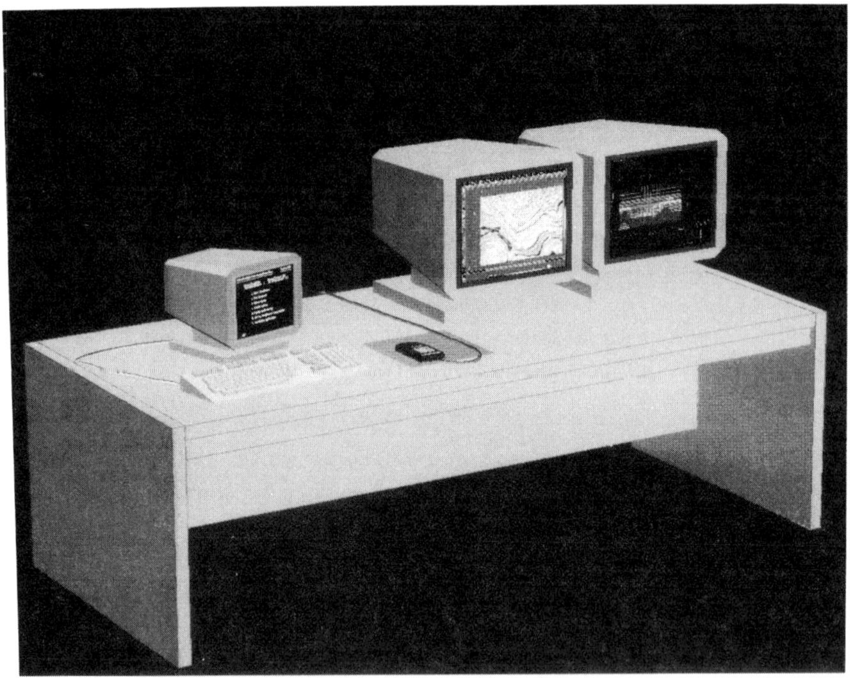

GeoQuest Systems, Inc.

Fig. 9–1 Workstation

have a set of prints of the inline sections, a set of the crosslines, and one of the time slices. With the data in triplicate like that, you'd have a hard time just getting your picks marked consistently on all those prints. And you still wouldn't be able to see a line from well to well, a flattened section, or a display of frequency or some other attribute, without ordering it from the processing center and then waiting for it to be made and delivered to you. With an interactive system, all those forms and more are immediately available, and any pick you make on one is automatically and correctly shown on any other form you display. A pick you make on any type of display—no matter what scale or proportion, attribute overlay, section, slice—can be automatically shown on any other type. Think of how hard it is to copy picks on paper to a different kind of section.

Working on a system is much faster than working by hand. When you mark a pick, there is automatic picking to take you rapidly over the easy parts. You don't have to stop and sharpen pencils, or make a pick lightly so you can erase it if you change your mind. And after you are confident of it, you don't have to heavy it up so it can be seen easily. Interactive picks are completely erasable. They are just computer memories of where you made the picks, and erasure is just telling the computer to forget that part of its memory, or change the magnetism on a disk.

The sections are in color. The color shows ordinary features clearly, and special color overlays can show seismic attributes on top of normal traces. You can select the colors to suit your eyes and your tastes, and change them any time you want to. Changing the colors can make features easier to detect.

In working on paper, timing picks, and plotting the times on a map are laborious, slow, and subject to error. With an interactive system, they require no effort, and are accomplished immediately and correctly. The system keeps track of your picks, and shows you where they are on intersecting lines, time slices, and maps. It is likely that a future generation of geophysicists will be almost unaware that those processes are going on.

You can analyze the data in many ways, quickly and thoroughly. You can make quick and accurate calculations on all the points you have picked on a horizon. You can change some of the parameters of a section that formerly could be changed only at data-processing centers. You can choose the scales of sections and maps, and have them produced at those scales immediately. You can change the proportion of horizontal to vertical scale of a section at will. You can change the overall scale to let you look closely at a detail or "back off" for an overall view.

Once you have picked a horizon on a section, you can flatten that horizon, showing the section more or less as of the time that formation was laid down. The flattening can also make it easier for you to pick other horizons.

The system makes maps for you and shows your picks on them. It can contour them and give you the opportunity to either accept its contouring or modify it. It can make 3-D displays of your interpretation (Fig. 9–2). You can have paper or film copies of any of the displays you can have on the screen.

SUMMARY

Summing up, interactive interpreting is better and faster than working on paper, but it takes more sophistication—computer, electric power, air conditioning. Which is better, the complicated system or a simple, reliable, slower way?

The power of interactive interpreting is so much greater that there isn't really any contest. Even with lack of portability and occasional times when "our computer is down", interpreters using interactive systems will increasingly find the oil that interpreters working on paper are still puzzling about.

In life in general, power methods of doing things tend to replace simpler, more basic ones.

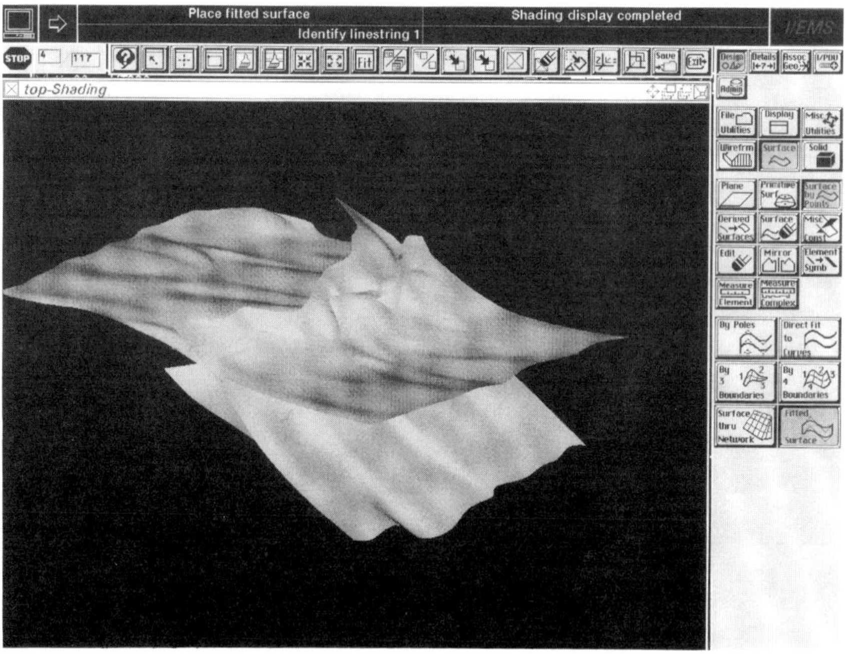

Intergraph Corporation

Fig. 9–2 Fault cutting through horizon

COMPATIBILITY

Interactive systems are changing rapidly, as the different organizations that develop them think up different new ways to make things work. That is characteristic of any rapidly advancing field. Standardization comes later, after some developments have been shown to be better, or at least more popular, than others. Railroads, for instance, were first designed with whatever width of track a developer thought would work well. But after trains became popular, it became necessary to link one set of tracks with another, so goods and passengers could be moved long distances without unloading and reloading on another train, one on a different gauge of track. The need caused people to settle on one gauge for a large area.

The acquisition and processing of seismic data have been around long enough for companies to have adopted standards, so the seismic data itself is usually compatible with all systems. The Society of Exploration Geophysicists has worked out some standards for the form of seismic data, the most popular of which is called "SEG Y". It prescribes the form to be used on magnetic tape for line number, shot point number, and other necessary information. Nearly every company uses the SEG standards in recording and in processing data. So when interactive systems were developed, they naturally were designed to use data that conformed to SEG standards.

Computers haven't reached that stage yet. There is talk of standardization, or at least compatibility, between computers, and some steps have been taken to achieve both. But we are still in the rapid-development stage, so compatibility is still a problem.

An interactive system must be compatible within itself. In selecting a system, if you like most of one system, but would like to include a feature from another source, the feature will have to be one that will work with the system. Or if you already have a system and would like to add a program or a peripheral, it must be compatible with the system.

Programs are written to work on certain computers, and usually won't work on others. Some programs have been "ported" to other kinds of computers, that is, rewritten to function with the other. And some programs are written to work with each other, and may not work with other programs you might want to use with them.

Incompatibility between systems can be a problem when you want to exchange data with another company in a joint operation or a trade, if your system is not the same as theirs. If your company has more than one kind of system, then that creates other problems.

The easiest way to have all of a company's systems compatible is to have them all be the same kind. But in real life, the resolve to do that breaks down. The company may standardize and then find that some special purpose can be served better with a different kind of system. A new system may come out with a feature that particularly suits one of the company's needs. Mergers and takeovers can bring into one organization equipment that had been acquired by two or more companies.

So one company comes to have a collection of systems. Interpreters have preferences among the systems, so it may be hard to have all the machines used efficiently. It is hard enough to learn one system well, without having to learn the commands for another. After you have learned a system, you are something of a specialist in it. Like learning foreign languages, the second one comes easier than the first, but, also like many languages, you can't just use your knowledge of one to get by in another. Even if you know two systems it can be confusing to switch from one to the other. The company may wind up with enough machines, but with some of them gathering dust while interpreters are lined up to work on others.

What does a company do with the machines that aren't used much? Try to get some use out of them? Buy updates for their software? Keep people maintaining them? Or let them sit near-idle until they are so out of date that they are worthless? It's a real problem for management.

NETWORKING

Like all computers, the computers of interactive seismic systems can be networked, that is, connected so they can transmit information from one to another by wire. If the different computers are compatible, networking them is easy. If the systems in a company are all connected, you can easily swap data with the other interpreters. Or, if you find that the work-station you have been using is busy, your interpretation can be passed by the network to another station and you can work on it there.

A network within a limited area, for instance within one building, is a LAN, Local Area Network. LANs are the most-used networks in seismic interpretation. The most primitive network, "sneakernet", consists of a person carrying the data on disk or tape from one computer to another.

WHERE DO THE IDEAS COME FROM?

Progress in geophysics comes about in various ways. There are practical ideas that occur to people working in geophysics. Some university research in recording and data processing has been supported by groups of oil companies. Some development in interactive systems is done by the research departments of large oil companies. And some is done by seismic contractors on their own.

But developing almost anything requires a bridging between two types of specialist—the people who make the things work and the people who use the things. In order to achieve that bridging for interactive interpretation systems, the tendency among makers of the systems is to gather information from their clients. They have meetings in which the clients suggest operations they would like for the systems to perform, ways they would like the operations to be easier to control, etc. Developers also ask clients to "beta-test" programs that are almost ready for distribution. The client uses the not-quite-final "beta" version of the program, looking for bugs in it. Some builders of systems even boast that their development is completely client-driven. It assures that the company won't spend a lot of time and money developing some feature that nobody wants. It gets features developed that interpreters want. It has the great advantage of making use of the experience of all the interpreters to improve the system.

But there is also another side to the situation. Sometimes people don't know they have a use for something until it has been developed. When the first personal computers were made, a survey of potential customers would probably have forecast a market consisting of only a few electronics enthusiasts. But when those computers became available, all sorts of people found uses for them. If some developer happens on a wonderful, but radically different, advancement in interactive interpretation, let's hope a customer survey doesn't discourage its development.

There is a growing tendency for companies to band together in a consortium to develop interactive interpretation technology. Several oil companies sponsor a contractor that does the actual development. The companies pay for the operation and provide guidance. There are meetings in which the companies describe the features they want the system to have, and review the contractor's progress. The system is developed for the use of the members of the consortium, of course, but at some time it is also offered for sale to anyone.

Why would companies handle the development this way? There can be several reasons.

- Although some contract companies develop systems for sale, oil companies would like faster development, or development along different lines, and want systems developed to fit their requirements.
- The project is expensive. If each company developed its own system, it would be spending a lot of time and money to work out a system that wouldn't necessarily be better than other systems.
- A system one company developed wouldn't remain secret anyway. As the company showed data to other companies for farmout and other deals, others would learn what the system did. Ex-employees would go to work for other companies. The mere knowledge of what functions a system performs is enough to allow a competitor to develop the same functions.
- It isn't even necessarily desirable for a company's system to be unique. A completely in-house system makes for clumsy communications with partners in projects, makes it necessary for new employees to be trained

in it, and doesn't let the company hire people who already have skills in the system.

WHEN DO YOU NEED A SYSTEM?

Changing from manual to interactive interpretation is a big step. In any major changeover, there are bound to be lost time, false starts, mistakes. This is costly to the company and is a setback, even though in this case the change will be a benefit in the long run. It is difficult to pick a time when you can afford to have things inefficient for a while.

The most sensible way to make a change like that is to plan ahead and work into it slowly. You could do that by getting one of the small systems, on which people could make analyses before getting a bigger system. Or you could get just one complete system, and at first train just one person to use it. Or the big system could be used a little by all the interpreters, so they could get accustomed to it at a comfortable rate. Any of those would be nice and orderly, but it isn't the way most human endeavors work. They are more often driven by crisis.

Companies tend to be driven into interactive interpretation by need. A company may make a discovery and need to shoot a 3-D survey over it before wells and other installations make it difficult and expensive to shoot there. Or the company may enter a deal that calls for its people to interpret a large 3-D area quickly. Not only is it costly in an interpreter's time, but even the paper prints of sections and slices, and the films from which the prints are made, can cost a significant fraction of the price of an interactive interpretation system. So the company is in a rush to get a system. That calls for several steps. People have to investigate the different systems and decide which one suits the immediate purpose as well as expected future needs. The system has to be bought or leased. Then interpreters have to learn to use it. And then the interpretation that caused all that activity can be made.

Probably better is the slower approach, that of convincing management—before there is an urgent need—that it is to the company's advantage to have a system, that a system will make interpretations more thorough, and that the better interpretations will lead to the company's finding more oil and drilling fewer dry holes.

CONVINCING MANAGEMENT

If you are an interpreter and think you could do your work faster and better with an interactive interpretation system, you need to convince management that it is financially worthwhile to get it. In comparison with other office equipment, the price is high, but it is cheaper than one dry hole. And compared with failure to make just one oil discovery, it is ridiculously cheap. If your company has a department that handles office equipment like chairs and copy machines, that group will throw up its hands. The system you ask for might take their entire budget for a year. The system is in a different category and requires that management get in the act directly.

There is a problem in presenting arguments to management. In persuading anyone of anything, it is natural to give the advantages more weight than the disadvantages, even when you are trying to be scrupulously honest.

When you tell the advantages of an interactive interpretation system, they include allowing you to work much faster, and helping you do your work better. People who are used to the way things are going may feel that you are interpreting fast enough (that's the normal speed) and well enough (you are finding oil). Possible rebuttals to your arguments might be that if you are too slow you should work faster, and that if you aren't interpreting as well as possible, you should try harder. You won't really be likely to hear those comments though, as most managers are accustomed to reasoning about the advantages of new tools.

One argument in favor of an interactive system is the illogical-sounding one, that other companies have them. That sounds at first like just trying to keep in style, but it is practical. If interactive interpretation does make it possible to find oil that a company not so equipped would miss, then there is a competitive advantage in having a system. Especially in working out a deal between companies, the company that obtains less from the data is at a disadvantage.

In convincing management, be careful that you don't raise expectations too high. It's easy to make interactive interpretation seem like magic, and then to bring on disappointment when it turns out to be something less.

MISCONCEPTIONS AND UNREALISTIC EXPECTATIONS

When you are getting a system, you have some ideas about what it will do for you. If you are an interpreter, vendors have probably given you demonstrations. If you are a manager, both the vendors and your interpreters have told you things it will do. But every advance has limitations as well as advantages. For instance, a typewriter produces words that are more uniform and legible than handwriting, but in typing, if you see that you are running out of room on the page, there isn't any way to make the letters smaller to crowd them in.

New users tend to expect that the promise of speed in interactive interpreting will be fulfilled right away. But first there is a learning period, before people can interpret rapidly. That can be a major disappointment if you got the system so you could meet an imminent deadline.

Demonstrations naturally make the interpreting look easy, with good data on the screen, and the sales person whipping the system through its paces. But in real use, you may have poor data. And you often won't go quickly through a series of steps, but will have to stop and make decisions. After you have finished interpreting an area, you too could rapidly go through it again, if there was any reason to.

It's easy to not realize how much preparation time must go into getting all your data into a system. The seismic data is probably in condition to be loaded fairly easily, but location data for combining lease information, well data, and different seismic projects may require a lot of time and effort to be put in uniform and compatible shape.

The speed with which you can interpret doesn't necessarily mean that you will complete an interpretation sooner. With the extra speed and other abilities, you will probably use some of the saved time to do a better interpretation.

And it still isn't magic. Even with the best of interpretations on the best of systems, you can drill dry holes.

BUY, LEASE, OR RENT

The best way to have a system available depends on your company's situation—whether the company already has a mainframe computer, how much 3-D interpretation it does, how much interpretation of any kind, how much money it is prepared to spend at the time, etc.

But if money was no concern, what would you do? When you think about it, that often is the case. That is, other considerations are more important than price. Balanced against interpretations not made or not made thoroughly enough, oil discovered or missed, wells and platforms correctly placed or mislocated, leases taken or not; the cost of the most expensive systems is small, even insignificant. But, of course, money is a concern, and you can't always afford to spend a lot now for some long-range advantage. If you are a consultant, your personal funds are limited. In a company, there are other people who have to be persuaded, and there too, funds may be limited at the moment.

Buying a system is a way to have it completely for your company's use. It makes it easy for company programmers to modify the system to suit company needs (if your organization is large enough to have its own programmers).

The hardware and software can be purchased from one supplier or different ones. The parts of a set obtained as a unit will all be compatible. But if the system is made up of parts from different sources, then someone has to make sure that the various elements, both hardware and software, will work together.

Or the equipment can be leased. An advantage of leasing is that it doesn't require as large an expense at one time as purchasing. And leasing makes it easier for the company to keep up to date. When hardware or software gets old, the company can lease newer forms.

Also, there are arrangements for the interpreter to go to the vendor's office and rent the setup by the hour. This can allow the companies to get some interpreting done without spending the time and money to buy or lease. It can also let the company's people try a system in real interpretation to see how they like it. It also makes it possible for a person consulting alone to use interactive systems. And being willing to use the system at night and on weekends may either reduce the price or, if the system is in great demand, make it possible to use it at all.

WHICH ONE FOR YOU?

If your company has decided to buy or lease a system, then you would like to get one with all the features you think would be most useful to you. But you probably won't be able to. It's like buying an automobile. If you make a survey of the features on different cars, you will be able to make a list of the features you would like on your car. But no one brand and model will have them all available as standard equipment, or even as options. Of course not. In a competitive business, each manufacturer is trying to think up and include good features that the competition doesn't have. Any feature that proves to be popular will be adopted in a year or so by the competition, but by then, there will be more new features.

Like shopping for a car, you might want to make a list of features in interactive systems, decide which ones are most important to you, and try to find a system that includes most of them. But don't decide to wait until one system comes along with all of them. Your competitors will become better at finding oil while you are waiting.

Some features, of course, are common to all the major systems. The systems all display sections and combinations of sections. They all let you make picks manually, automatically, or in between. But there is a great range of abilities in some other features.

Where do you get up-to-date information on the different ones, to put in your list? A book won't give you the very newest information, because the time it takes to write it and then to publish it is long enough for the systems to change radically.

Magazines are published more quickly than books, and so can give you information that is more up-to-date. Two magazines that may have comparisons of systems are the Denver Geophysical Society's *The Record* and the *Geobyte* of the American Association of Petroleum Geologists. Both have published reports of the Denver society's C.E.E.D. (Comparison, Evaluation, Exhibition, and Demonstration), in which various systems ran standard tests. It is hard to design tests that measure performance in real interpretation, though, so you will also want other knowledge of the systems. The publications of other local societies may have useful information about seismic workstations.

Looking at advertising, both in magazines and in brochures handed out at conventions, is a good first step in comparing systems. The ads can be found in the Society of Exploration Geophysicists magazines *Geophysics* and *The Leading Edge*, the American Association of Petroleum Geologists *Explorer* and *Bulletin*, and *The Oil & Gas Journal* and *World Oil*. That part's easy, just look at the ads in the magazines. Some of them explain technical features well. Ads only show the good side, naturally, but they do give a fair idea of what features a system has. And they even, in a way, mention problems. If an ad claims that its product solves some problem, you know there is such a problem. And even the solution mentioned may not completely resolve it. You will need to watch out for that problem in all products of that type.

One way to get more of the feel of the systems is to attend meetings like the SEG's annual and regional conventions and the Geotech meetings. But a demonstration at a meeting is necessarily brief and general. Although the exhibitors are glad to answer your questions, meetings aren't good places for getting thorough demonstrations. There are too many other potential customers there, and the exhibitors can't neglect them. Your company can get a more thorough demonstration of a system it is interested in, at some other time, by requesting it from the vendor.

Another source of information on systems is from other users. They too can't tell you how well a system will fit your purposes, but they can give you an idea of the problems they have with it. That is sort of the reverse of what the advertising gives you, so the two together may sum up a system's advantages and disadvantages pretty well.

If you happen to have occasion to rent time on systems to interpret areas, you would be able to compare them well. Of course, using several systems one after the other, and learning new commands with each one, would make your interpretation less efficient than if you could learn one system well. But that might not be too high a price to pay for being able to compare systems in the light of your own experience.

But without that varied experience, how do you decide which system is best for you? You probably can't, because it depends on so many factors, some of them imponderable. For instance, after you install a system, will the company that made it continue to make improvements that you want, or will some other system take the lead in a direction that you then decide you like better? And you can't even decide how well a system will work for you and your purposes until you use it for a considerable time. Even the renting of time on different systems may not give you enough experience on each one.

In making this decision, all you can do is to get a pretty good idea of which system comes nearest to fitting your needs. Choose a system carefully, but after you have it, quit thinking about other systems, and concentrate on learning to use it as well as you can.

AFTER THE SELECTION

When you acquire a system, you start receiving support from the vendor. Manuals are supplied with the system. Training courses for several of your people may be included. There will probably be a "hot line" service, assistance over the phone when you have problems. And some vendors organize user meetings in which the customers get together with them and make criticisms and suggestions for improved or additional functions. You will not be left alone.

When you have a system and start using it, you are committed to it, just like you are to a house or a car you buy. That isn't a disadvantage, it's more of an advantage. As you use it, you will learn its quirks, learn to take advantage of its good points, find ways around some of its weaknesses, get the best out of it.

There are two ways to learn to use a device: the "school way" of learning all about its background and fundamentals, and the "cookbook way" of learning to use it by following simple instructions. The school way gives you a more solid foundation in it, but you have to complete the whole course before you can start getting practical use out of it. The cookbook way doesn't give you any background information, but lets you get started quickly, and then gain abilities while using it.

Unfortunately, most computer manuals are written in the school way, so you need to learn the whole manual before being able to use the system. One reason they are written that way is that computer programmers understand programs from the development viewpoint, and when they are called on to write instructions, they tend to write them from that direction. So an instruction may explain what a command causes the computer to do. But an interpreter needs to know how to get some particular interpreting operation performed.

The instructions for seismic systems aren't completely programmer-written, though. Companies make efforts to write the instructions more for the interpreter's use, and are probably at various stages in that transition, from instructions that tell what the programmer has done, to instructions that tell the interpreter how to do something.

To really learn the system by studying it in manuals would take a lot of time during which you wouldn't be productive in interpretation. So the usual way to learn is by being taught to use the main commands. You are taught in the cookbook manner, taught to do this, and then do that.

The makers of systems do provide good cookbook training, with their people showing you what commands to use, starting with the simpler parts of the system. After a few sessions of that kind of teaching, you start using the system on your own. After that, you will ask people questions and work some commands out. One good way to do is to just try things. You can get hints from the names of the commands—they were designed to give an idea of what they do. That's part of the fun of computers, seeing what you can do with them without help.

Familiarity will make it yours, and the skill you gain will make it increasingly useful to you. If there are features you come to wish it had, talk to the manufacturers; they are all eager for user input. But don't spend time wishing for another system. The others aren't perfect either, and you aren't familiar with them. This one may well become the best one for you, because you know it intimately. The more you learn it, the more tricks of using it you work out, the better it is for your interpretation.

OUTWIT THE COMPUTER

Computers do a lot of things, but those things are all rigidly defined. If you want to do something just a little different, the program may not be able to do that.

A problem with many systems is in annotating. Some systems have annotation functions to label a section, maybe one that you want to show people as an example of some phenomenon. They allow you to type words, draw arrows and other symbols. If your system doesn't have such a function, you may be able to annotate with the well-labeling function. Tell the computer you want to put a well at a certain shot point, give zero for the total depth of the well, so there won't be a vertical line to represent the well bore, and use the command that is meant for adding formation names. Then type your note—the computer doesn't care what you name a formation.

If the system does have a command for annotating, though, you still may want more. The letters and lines may not have the bold look you may want. The lettering is neatly horizontal, maybe in fairly small characters, and characters and everything may be formed from thin lines. If you want to mark the section in a more flamboyant way, with big letters, thick lines, bold arrows, words at an angle, the function won't do it.

To get around that problem, you can give the command to start a new horizon or fault. If your system allows only one pick to be made on a horizon at a trace position, a horizon won't do. But some systems allow as many picks as you want at a point on a horizon, mostly so you can pick a horizon's two layers at an overthrust fault. If your system accepts only one pick per trace for a horizon, you may be able to get around it by telling the system that you are picking a fault; the systems usually allow more than one pick per trace for a fault. They allow that because fault picks may be kept all under one name until you can decide which fault crossings you want connected from line to line as a single fault, and because a near-vertical fault may waver back and forth.

Start a horizon or a fault, whichever works better, give it a name, go into the freehand picking mode, and "pick" in hand lettering, longhand, drawings, arrows, whatever you want. You can choose a bright color for the "horizon", and, if the system lets you choose line widths for picks, select a fairly thick line for it. Then the fake horizon can include any bold-looking marks you want. Later, working on the section, when you don't want your sign in the way, just don't call up that "horizon" or "fault".

This technique allows you to outwit the computer to solve another problem, the problem of marking up a section at the start of work on an area. You may have viewed a movie or sequence of displays to prevent big surprises of a general kind in working the area. After that, to get an idea of what you are likely to encounter in the finer points of actual picking, you may want to call up a typical section (after viewing the movie, you know what is typical) and pick a lot of reflections on it. In this early look, you are not concerned with identifications, you are just looking at the pickability of the data, and the forms of reflecting horizons. With paper sections, many people start with a yellow pencil, and just mark all the reflections they see. Why people choose yellow, I don't know. I used to think I was the only one who did. Maybe it is popular for this purpose because it is a color that isn't likely to be chosen for a "real" horizon, one you are going to pick all over the area, and then map. It's a nice subdued color, not standing out like red, blue, or green. Of course, the particular color doesn't really matter.

If your interactive system accepts only one pick per trace for a horizon, then any time you make a pick above or below another one, the system considers it a correction, and deletes the first pick. So tell the system you are picking a fault or a horizon, whatever works on your system, and then go ahead and make your yellow, or whatever color you want, picks all over the section.

Once you have picked a section pretty fully, you can sit back and look for interesting features on it—they show up better with the alignments marked. It may make depositional features clear, with different slopes for onlapping deposition and deltas, etc. It may show you where reflection quality is poor; it may make fault breaks easy to see.

You probably won't want those picks displayed on the section later when you are making specific picks, so you can give this "fault" or "horizon" a name like "Trial" or "Test" or something that implies to you what it is. Then just don't display it when you are doing regular picking. It's still available any time you want to look at it again. Or you can set the picked section aside as a picture, in the special memory for things that are to be displayed but not worked on, or made into hardcopy. After saving it as a picture, you can delete all the picks from the section, or cancel the "fault" or "horizon".

In another example of outwitting, suppose you want to adjust a section to agree with well data on a shallow formation. That horizon is easy to pick on the section, but it looks irregular, with ups and downs that you are reasonably sure were caused by near-surface problems, like weathering, permafrost, tides. From well and outcrop data outside the area, you know the regional dip of the formation, and believe that the dip is uniform, without the irregularities the section shows. After picking the horizon, you can easily flatten it by just giving a command. That smooths the other reflections on the section so they are in the positions they were in when the shallow one was deposited. Fine, that's a good, useful thing to do with the data. But you want to see how they are now, not distorted by the near-surface problems, and not distorted by the flattening of the shallow horizon. You want it to dip as you know it now does.

What you need is to unflatten in a special way, introducing a tilt to the section. Putting in the tilt seems simple enough, a lot simpler than the flattening was. But on your system there isn't a command to do that. It wouldn't have been hard to add, but the need wasn't foreseen. What can you do?

The system has the ability to perform some calculations. Maybe they will help. You can draw a line to represent the shallow horizon, by using the rubber band function all the way across the section. That produces a long, tilted, straight-line "pick". That fake horizon looks like any other horizon to the computer. But it doesn't do anything for your other horizons. Aha! Here it is—you've already flattened the shallow reflection. On that flattened section, put your straight-line tilted horizon, or just imagine it, you won't be doing anything with it. Then pick another straight-line "horizon", with the opposite tilt. Flatten this new, reverse-dip horizon. Behold! The first horizon you had flattened is tilted just the way you wanted. And all the other horizons are now in their correct configurations, to the degree that the tilt you obtained from the well data is correct.

These are just samples. As you get familiar with a system, you will find many ways to "fool" it into performing functions the programmers didn't intend.

KEEP YOUR EQUIPMENT UP TO DATE?

After getting your equipment and learning to use it and also to outwit it, what do you do about keeping the system up to date with the passing years?

One method of keeping up to date that I mentioned is to not own the equipment at all, but instead to lease it, a year at a time. Then, when equipment you are using becomes obsolete, you can lease more modern equipment. The lease contract may specify that equipment will be replaced as it becomes outdated. Of course, for either of these plans to work on a continuing basis, the price for leasing has to be high enough to pay for the repeated updating.

If, instead of buying or leasing, you choose to rent time on a system, then you don't have this decision to make, since the vendors try to have their latest models available for you.

If you have purchased the equipment, the vendor will have some arrangement for keeping the equipment working and for updating it and the software. The arrangement may be in the form of a maintenance contract, like the service contracts people sometimes get on major household appliances. With changed programs, though, there are necessarily some new commands to be learned. So you will have to be prepared to learn new techniques. That will take some of your time, but the improvements will probably save time in the long run or give you new capabilities. That's not very different from any kind of work that is affected by technical progress.

Another effect of progress is in the way a program can get more complicated as new features are thought up and added to it. A program is originally written as a unit, with the various characteristics planned in relation to one another. But a new feature, being an add-on, can't be as well integrated with the rest of the program. After a number of features are added, the program becomes more complicated, with more additions that don't fit neatly into it. It begins to take longer to run than it would if it had been a new program with all those features integrated.

There are some ways that the hardware is constantly improving. These changes are to more memory, more speed, more power, smaller size, and lower price. They would keep you hurrying if you tried to keep up. In these respects it is probably better to catch up occasionally, using a system long enough to be comfortable with it for a while, and then making a change to newer equipment.

So of course the answer to the question of whether you should keep your equipment up to date is that eventually you will get a new system, more powerful and maybe cheaper than the one you have. Meanwhile, to decide when to do that, you need to consider ways to keep the operator—you—aware of new developments.

KEEP YOURSELF UP TO DATE

For years there have been rapid changes in both shooting and processing of seismic data. But until recently, interpretation had failed to improve much, either in the way people did it or in the equipment they used. Now, though, interpretation is advancing at a breakneck pace. This book is mostly about deskSIZE computer setups and terminals attached to powerful computers, but deskTOP computers are working their way into interactive interpretation. Even laptop computers can already be used to some extent. Achievement of the goal of having a system available for every interpreter is nearing.

How are you to keep up to date? Mostly, you can't keep right up to the latest improvements. But you can keep yourself aware of the changes, especially after they arrive on the market. How do you do this? You will stay up to date more or less the same way I mentioned in the topic, "Which One for You?" (by reading the ads and articles, attending professional meetings, and talking to people). For practical information on advances in interactive interpretation you can read articles in *The Leading Edge, The Oil & Gas Journal,* and *World Oil* and publications of regional societies. There are also newsletters about various related topics. One such, *The Pohlman Report,* is devoted to developments in the various interactive interpretation systems.

Both in connection with the conventions and at other times, there are courses offered in many aspects of seismic interpretation. Some of the courses are given by companies, some by individuals, and some by the societies.

Sometimes you may be able to get on a mailing list for a company magazine. These magazines often have informative articles on new developments.

You don't have to be fanatical about keeping up with the latest style in interactive interpretation. The leap you make in changing from interpreting on paper to using an interactive system is the big one, much bigger than a change to a more advanced system. But it does pay you to keep alert for new developments in the systems and new ways of using them.

WHAT'S NEXT?

In gathering information for this book, I went to a geophysical convention, concentrating on vendors' exhibits and on the demonstrations the companies gave. Walking down the aisles between the exhibits, and then sitting down watching demonstrations, a feeling of wonder came over me, wonder at the power these computer systems give interpreters.

At workstations, we pick and map far faster than we ever could before. We analyze data immediately instead of worrying about whether to spend the time and money to have it analyzed. We make color displays just to see what they might show, displays that would once have meant more time and money. We vary the playback parameters of sections at will.

Those are things we can do now (Fig. 9–3). At the convention I experienced another feeling of amazement when I thought about the implications for other things that we will be able to do in the future, not the far future but next year or even next month. Progress in interactive interpretation is rapid and continuing. I get out my pocket calculator—my cheap pocket calculator—and consider that it probably does as much as the first computers, the ones that filled rooms and had technicians running around replacing vacuum tubes. Those first ones were made between just forty and fifty years ago. Progress in computing has accelerated recently. What will we be doing just ten years from now? The slang expression "It boggles the mind" is appropriate.

I keep saying that the field of interactive interpretation is advancing rapidly, and keep trying to give the most recent advances. But since it is moving so fast, there are other changes to come, some soon, others farther into the future. You might want to know what some of the coming developments are.

The best source of information on coming improvements in interactive interpretation is the magazine *Geophysics*. It features the things the researchers are working on now, things that will be the practical aids to finding oil in the future. At least, some of them will; some will turn out to be dead ends.

Reading *Geophysics* is easiest if you have just completed a calculus course. There are many integral signs and other abstruse mathematical expressions in the articles. But, if you aren't trained in math, or no longer remember much of it, you can still get information from the articles. Each has an abstract, and almost every article has an introduction and a conclusion. Those three parts are free of math. And skimming the body of the article, skipping over the equations, will yield some information. Many of the articles are about details of seismic processing, but a few will lead to advances in interpretation.

The same kind of information can come from geophysical conventions. The papers given at these meetings are often the same ones that appear later in *Geophysics*. The talks may be somewhat less technical though, because the math is necessarily curtailed for a presentation.

That's for the specific advances, mostly small changes in interactive interpretation. But in the long run, in general, what can be expected?

Present trends will certainly continue. A sudden stoppage of progress is difficult to imagine. Research on principles, and development of products, are always underway. They are at any time at many different stages, from remote to almost ready to go on the market. They would all have to stop, the most developed ones immediately, for progress in interactive interpretation to come to a halt. **There will continue to be progress. The systems will continue to change.**

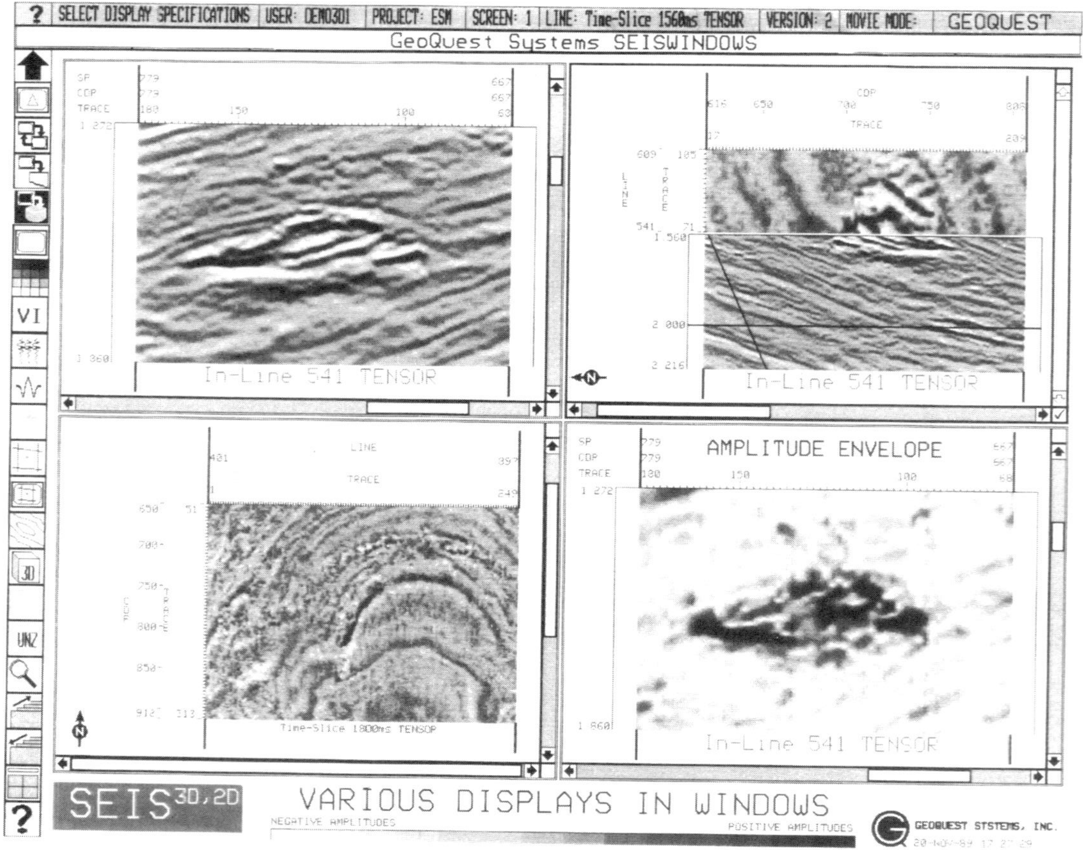

GeoQuest Systems, Inc.

Fig. 9–3 Some capabilities of interactive interpretation

Computers are getting more powerful; they are also getting smaller and cheaper. Notice that I didn't say that they will be powerful or small, powerful or cheap, but powerful and small, powerful and cheap.

Power is the ability to handle complex arithmetic operations rapidly. This too will increase. We can expect desktop systems that do more than the largest mainframe systems do now.

Compactness is useful in interactive systems. If a system occupies a desk top, it uses less floor space than one that is spread over two rooms. Floor space is leased by the square foot, so a smaller system costs a company less in rental for space. A small system may be more easily moved from one room to another, or one city to another. Going further, one that is hand-portable, with the same capabilities, would be more useful than one that has to be moved like moving furniture. Computers will get smaller, and so will interactive systems.

Price, too, will continue to change. Since computers were invented, they have become cheaper while increasing their power. Lower prices for interactive systems mean that more of them can be available to interpreters and that people won't have to wait so long to use them. Eventually, powerful systems will be so cheap that each interpreter can have one to be used whenever needed. That will mean that there won't be any economic disadvantage in having a system sit idle, any more than there is now for a typewriter. And, perversely, when that is true, the systems will be used for more interpreting functions than they are now. If you have a system at hand, you can use it for all sorts of things, whenever they occur to you. The systems then will be the best of servants, unobtrusive but always standing by ready to do your bidding. Fortunately, this reduction in price will enable you to more easily get approval for new systems that have larger memory, more speed, and more power. Eventually we may have systems that are cheap enough for each interpreter to own one personally, like people now own pens instead of depending on the company to furnish them.

There are some other characteristics of computers that will also continue to improve: memory, speed of operation, and display.

A computer needs a large memory. But computers don't ever seem to have enough memory. If a system has more memory than it needs, then developers will find ways to use the extra memory to make the system accomplish more. And if they don't, then users will decide to load more data into the system to make more comparisons, or to work larger areas. By a sort of law of nature, or at least a generalization about how people use what is available to them, you will probably never, for very long at a time, have all the memory you want. It's like money. Can you imagine having so much money that you wouldn't ever work up a desire for something you couldn't afford? Systems will continue to have more and more memory, but it will never seem to be too much, and it will rarely even seem to be quite enough.

The same goes for speed of operation. Computers allow people to work much faster than they could before, but people get impatient if, at any stage of this faster operation, they have to wait while the computer performs some task. And when the speed increases, that allows the computer to be useful for some new function, one that before would have taken too long to be worthwhile. The chase after greater speed will also continue.

The improvements just discussed are trends and therefore easy to foretell, but what about some new changes that are needed? Well, I've been saying that a problem with present-day systems is that their screens are too small. Future systems, even though smaller than present-day systems, will probably have larger screens, not smaller. There are two ways that could go.

First, the screen could be thin and flat. There are some such screens in use now on laptop computers. They aren't nearly good enough for interactive interpretation yet, but we are talking about future developments. Since they exist, we know there will be improvements in them—there are improvements in everything. A screen for a future small system might be one that could be hung on the wall, or laid flat on a desk like a paper section. And, of course, the real problem about screen size, the amount of detail, the number of pixels that can be put on the screen, will also increase. You might handle a big flat screen something like a drafting board, tilting and turning it to a convenient position.

The second way of handling displays for smaller systems has also already been invented, and only needs improvement. It uses a sort of goggle that projects the image directly to your eye, so it appears to hang in the air in front of you. You see whatever is in the goggle. If a section is in it, then there is no need for a wall or desk to put it on. All the seeing you do now depends on light that reaches your eye, anyway, so the image might as well go directly there instead of onto some surface you point your eye at. With this method too the detail depends on the number of pixels that can be incorporated in the image, but with development they will increase.

For instance, with more pixels, the system can show a highly compressed section just by squeezing traces together, without having to leave out any of the traces. But for it to show more detail in a trace, there will also have to be an increase in the sample rate in data collection in the field. This too can be expected as recording instruments become able to handle more data.

What advances should there be in the functions a system performs? Well, automatic picking will surely become more sophisticated, so it does almost all the structural interpretation for you. And with that done, there can be automatic production of structure-based displays. You could call up a wire-frame display of a horizon and it would be picked and displayed immediately, or you could similarly ask for a horizon slice and the horizon would be picked, flattened, and displayed for you. Attribute displays will become more advanced, showing other characteristics of the data that people haven't yet thought of.

We are just starting to use amplitude-versus-offset displays in section form. That will surely get more sophisticated and more useful, being shown in the form of sections. And attributes other than amplitude can also be displayed to show how they vary with different offsets.

We have also just started using seismic techniques to measure changes in the subsurface that take place over months and years, for instance the advance of a waterflood secondary-recovery process. Extending this idea, you can imagine a 3-D survey, using permanently buried geophones (as are already being used to some extent), conducted on a regular schedule to monitor oil production through the life of a field.

The role of the interpreter will change. As noted in Chapter 1, the interpreter is already doing things that were formerly done only in data-processing centers. With so much of 3-D shooting being done for field development, the interpreter will necessarily work more with engineers and development geologists. The systems will add functions for this work. Other functions will be added specifically for seismic stratigraphy. As VSPs become more common, interactive systems will have functions especially suited to interpreting them. As more shear-wave data is gathered, there will be ways of taking advantage of its special qualities. Shear-wave sections will become common. As seismic tomography is further developed, it will bring about changes in interactive interpretation systems.

Do all those additions to the abilities of interactive systems mean that you will become an expert in processing, seismic stratigraphy, AVO, etc.? Not likely—you will have all those abilities at your fingertips, and can use them when you want to, but you will naturally tend to specialize in some of them. Instead of making every interpreter a generalist, the added abilities may divide seismic interpretation into more and narrower specialties. Already there are companies that specialize in loading data, companies that analyze attribute displays, others that make intensive studies of reservoirs. Those specialties didn't exist a few years ago.

How nearly correct are these predictions? In detail, probably pretty good; in general, not so good. The problem with predicting technical advances doesn't seem to be with whether a device will be invented—if it's at all reasonable it probably will be—but with the timing. If I correctly predict four devices, but don't have much idea which will be invented first, then I don't have much idea of what some future period will be like. The first one to be developed changes the world. At some time after that, two of them are present, but the other two haven't come along yet. And one of the early ones may change the way a later one develops.

Will all this automatic stuff put you out of business as an interpreter? Of course not, it will just make you able to spend your time with more subtle aspects of the subsurface, so you can find more oil and produce it more efficiently. It will also help you find, and probably help to produce, other minerals in the earth. Seismic methods have already been used in looking for coal, potash, and gold. The more you know about the subsurface, the more you can accomplish with that information.

AT THE WORKSTATION

You are sitting at the workstation, just finishing a session. With several interpreting sessions completed, you are now familiar with many of the aspects of your system. There is competition for time on the machine, so you have chosen to work at night. It's pleasant, there aren't many interruptions. The few times the phone rang, you weren't expecting any calls, so you either didn't answer it, or if it was persistent, you told the caller to try again during office hours. You have become fairly well acquainted with the building's cleaning staff.

The sun came up some time ago. You like this kind of work, but are tired and ready to go home. You exit from the program, and the computer records the time you finish. Daytime people start straggling in. You've had the whole place to yourself so much that it almost feels like they are invading your territory, using the coffee machine and the desks and chairs around the office. Your boss comes through the door. "How was it?"

You're ready to leave, but on your schedule there aren't many chances to talk to the boss about your work. "There is this one interesting feature that I hadn't noticed until last night".

You reach over and call up the program. . . .

Index

A

B

D

E

F

G

H

O

P